STEAM TURBINES

A PRACTICAL WORK ON THE DEVELOPMENT, ADVANTAGES, AND DISADVANTAGES OF THE STEAM TURBINE; THE DESIGN, SELECTION, OPERATION, AND MAINTENANCE OF STEAM TURBINE AND TURBO-GENERATOR PLANTS

AF825254

WALTER S. LELAND, S. B.

FORMERLY ASSISTANT PROFESSOR OF NAVAL ARCHITECTURE,
MASSACHUSETTS INSTITUTE OF TECHNOLOGY
AMERICAN SOCIETY OF NAVAL ARCHITECTS AND MARINE ENGINEERS

ILLUSTRATED

(815) 935-5353

Steam Turbines

by Walter S. Leland

Copyright © 1984 by Lindsay Publications, Bradley, IL
Original copyright 1910, 1917 by American Technical Society, Chicago, Illinois.

All rights reserved. No part of this book may be reproduced in any form or by any means without written permission from the publisher.

ISBN 0-917914-20-1

1 2 3 4 5 6 7 8 9 0

INTRODUCTION

THE steam turbine is not only one of the most ancient forms of prime-mover, but is also one of the most recent engineering developments, and perhaps the most talked-of at the present time. Remarkable advancements in its development have been made during the past ten years, and this has given a great incentive to steam engineering, while the electrical engineer has been stimulated to work out new ideas in designing generators suitable for the very high speeds of rotation. The steam turbine has forged rapidly to the front, and in spite of the early and serious handicaps, in the way of steam economy, has taken its place beside the best reciprocating engines of the present day. The turbine possesses many advantages over the reciprocating engine, and its field of greatest influence is likely to find in the immediate future a more serious competitor in the gas engine than in the reciprocating engine.

¶ With the relative merits or shortcomings of the steam turbine and the steam engine as prime-movers, the pages following are concerned to a certain extent, for modern engineering has not yet defined with sufficient clearness the respective fields of these agents, into either of which the other agent may enter with a certain degree of success. The rivalry between these two forms of prime-mover has, however, given place to the mutual recognition of the merits of both, and cordial co-operation in certain classes of work for which neither alone is fully adapted; as, for instance, the use of the low-pressure steam turbine with the reciprocating engine, in which a higher economy is effected than that obtained by the use of either alone.

¶ Steam turbine developments of recent years have been in the direction of higher economies in the production of power from steam, and while the ultimate limit has not yet been reached along this line, still the steam turbines of today are highly efficient machines, measuring up to the best energy economy.

¶ The pages which follow have been written for the practical man, for the man less interested in the finer points of theory than in the results accomplished and the way they are secured by the most successful builders of the steam turbine. Nevertheless, the fundamental principles have been brought out in sufficient detail for intelligent mechanical design. Between the designs of different builders, little attempt has been made to draw comparisons; but the facts have been stated with the idea that the reader may form his own conclusions on points where a difference of opinion may arise.

TWO 1875-K.V.A. THREE-PHASE 11,000-VOLT LOW PRESSURE UNITS. INSTALLED FOR THE DUQUESNE LIGHT COMPANY AT RANKIN, PENNSYLVANIA

Courtesy of Westinghouse Machine Company

CONTENTS

CONTENTS

WARNING

Remember that the materials and methods described here are from another era. Workers were less safety conscious then, and some methods may be downright dangerous. Be careful! Use good solid judgement in your work, and think ahead. Lindsay Publications has not tested these methods and materials and does not endorse them. Our job is merely to pass along to you information from another era. Safety is your responsibility.

Write for a catalog or other unusual books available from:

Lindsay Publications
PO Box 12
Bradley, IL 60915-0012

5000-K.W. CURTIS STEAM TURBINE

Courtesy of General Electric Company, Schenectady, New York

STEAM TURBINES

PART I

Introduction. The steam turbine is one of the most recent engineering developments, and perhaps the most talked of, at the present time. During the past ten years the most marked improvements in its development have been made, and this has given a great impetus to engineering, especially steam engineering, although the very high speeds of rotation have driven the electrical engineer to work out new ideas in designing generators suitable for these higher speeds. The turbine has forged rapidly to the front, and, in spite of an early and serious handicap in the way of steam economy, has taken its place beside the best reciprocating engines of the present time. Many claim it to be superior in the matter of steam economy, but this will be discussed more fully later on. The turbine evidently possesses many advantages over the reciprocating engine, and, in its field of greatest usefulness, is likely to find in the near future a more severe competitor in the gas engine than in the reciprocating engine. For some classes of work, the steam turbine in its present state of development is entirely unadapted.

The steam turbine consists essentially of nozzles or guide passages which direct the steam onto vanes or buckets attached to the periphery of rotating wheels, the essential elements of which are shown in Fig. 1. The simplest form of turbine is perhaps one of the type in which a jet of steam impinges upon the buckets of a wheel, in much the same manner that a stream of water impinges upon the buckets of a Pelton water-wheel; there is, in fact, a great similarity between water turbines and steam turbines. The underlying principles are the same in either case, but the application of those principles is different. Steam flowing through a properly designed nozzle, with 150 pounds boiler pressure on one side, and the usual turbine vacuum on the other, will attain a velocity of about 4,000 feet per second, or about twice the muzzle velocity of a rifle ball. Water, to attain this enormous velocity, would have to flow from a head of about 234,000 feet. When this is compared with the ordinary water head of 150 feet or less, or even with such an exceptionally high head as 3,000 feet, which is sometimes met with in water powers on the Pacific Slope, a glimpse will be had of the magnitude of the problem confronting the steam turbine engineer. To put this in other words, the steam turbine designer has to deal with a velocity equivalent to that produced by a head of water nearly

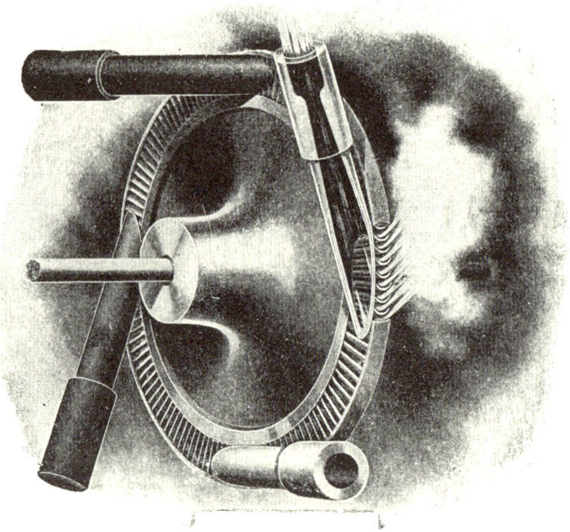

Fig. 1. Elements of DeLaval Turbine.

1,500 times as high as Niagara Falls. It will at once be seen, then, that the velocity of rotation of a simple turbine wheel to attain the best efficiency must be enormous. If this total head is to be used in one wheel, the peripheral speed must be nearly 2,000 feet per second, and at such speeds, the centrifugal force is so great that it is no easy matter to design a wheel that will not burst, even were there available some material stronger than any we now know. As it is, about 1,200 feet per second is considered the practical limit of peripheral velocity for a wheel built of the best nickel steel.

A little mathematical calculation will show that wheels of five feet in diameter will revolve 4,600 times per minute to attain a velocity of even 1,200 feet per second at the periphery. It is the problem of the steam turbine designer to reduce these speeds to more manageable rates without at the same time making too great a sacrifice of efficiency.

From a thermodynamic standpoint, the turbine and reciprocating engine are not unlike, but the force of the steam acts differently in them. In both, it is the heat energy of the steam that does the work. In the one, the steam slowly expands, exerting pressure on a piston; in the other, it expands in narrow passages, pushing the particles ahead faster and faster and thus obtaining velocity which is then imparted to the vanes of a rotating wheel. In the one, the steam acts by virtue of statical pressure; in the other, by virtue of its high velocity. In either case, it is the internal heat of the steam that causes the expansion and does the work. If heat is lost in any way, by condensation, radiation, etc., the work will be proportionally less. In the turbine, a difference in pressure from inlet to outlet acts as a motive force indirectly, and then only in so far as it causes a rapid flow of steam.

Advantages. The well-known expression, *work* = *force* × *space*, embodies the idea that a given amount of work may be accomplished in a certain time by increasing the total force of the steam on the piston at the expense of the number of revolutions of the fly-wheel per minute, or *vice versâ*. For example, a Corliss engine running at 180 revolutions per minute requires a mighty effort behind the piston to develop 1,000 horse-power, and this tremendous force demands a large cylinder, a heavy frame, and an immense fly-wheel. If, now, an engine were built to run at 800 revolutions per minute, much less push behind the piston would be necessary to develop the same horse-power, and, therefore, the parts could all be made smaller, and the whole weight very much reduced.

To go a step further, and consider the steam turbine, which must run at 2,000 to 3,000 revolutions per minute, it is clearly seen that this enormous speed reduces the mass of the parts even more. The heavy fly-wheel is no longer necessary, as the rotating parts are moving at a sufficiently high speed to acquire an immense inertia, and there is always a constant effort exerted on the vanes by the team, thereby producing an absolutely steady turning moment. Furthermore, the motive parts of the turbine revolve, which is in direct contrast to the reciprocating engine, in which the piston is moving backward and forward, and the turning moment is continually changing from a maximum to a minimum. It is clear, therefore, that for a given horse-power, the steam turbines produce smaller machines, lighter foundations, and consequently smaller power houses. A fair idea of the relative space occupied may be gained from Fig. 2.

Again, the generator, on account of this high speed, will be smaller and less expensive. The turbine requires oil in its bearings

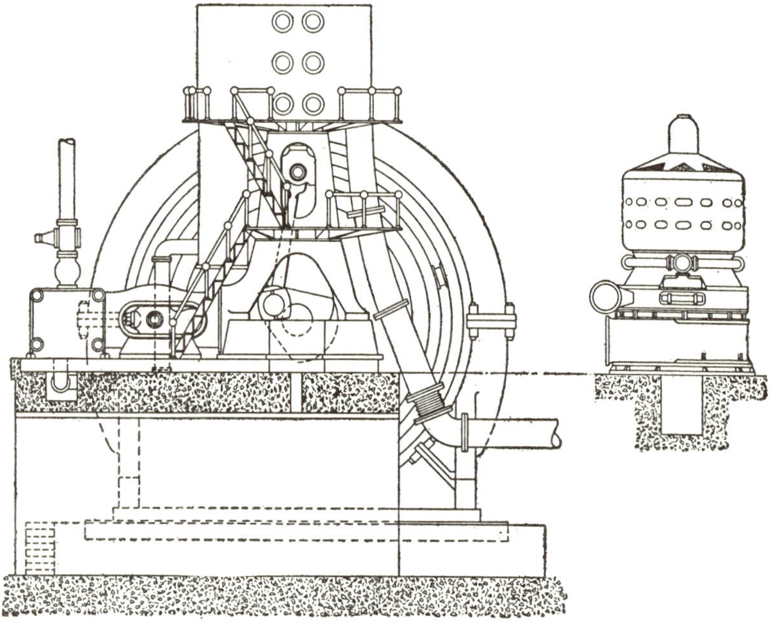

Fig. 2. Comparative Sizes of 5000-K. W. Corliss Engine and Generator and Curtis Turbo-Generator of Same Power.

only; hence there is no oil to go over in the condensed steam, and the condensation may be used for boiler feed without any danger of carrying oil into the boilers. The turbine requires somewhat less attendance than the reciprocating engine, and the whole machine is compact and simple. To do its best, the turbine requires a higher vacuum than is ordinarily obtained for the reciprocating engine, and hence needs very much larger condensers, more cooling water, and additional air-pump capacity. All this in a measure offsets some of its advantages, and frequently more trouble arises from the air pumps and condensers than from the turbine itself. The turbine may of course operate at the usual vacuum with a somewhat greater steam consumption and a slightly lower efficiency.

The reciprocating engine has its own advantages, and in certain classes of work will doubtless hold its own, but for all such apparatus as blowers, centrifugal pumps, generators, etc., which may be direct connected to a turbine, the reciprocating engine is rapidly becoming a thing of the past, and even for factories where belt drives are used, the steam turbine has been suggested.

History. The steam turbine is not only one of the most recent engineering developments, but is, at the same time, perhaps, one of the most ancient forms of prime mover. In a book written by Hero of Alexandria, over 100 years before the beginning of the Christian era, a very simple form of steam turbine is described. It consisted of a hollow sphere mounted on hollow trunions, through which the steam passed into the sphere. On opposite sides of the sphere were outlets consisting of pipes bent at right angles in lines tangent to the equator of the sphere, in such a manner that the reaction of steam escaping through these pipes caused the sphere to revolve on its trunions, in much the same way that the water escaping from the arms of a lawn sprinkler causes it to revolve. This turbine, which is illustrated in Fig. 3, is the simplest form of the pure reaction motor.

Fig. 3. Hero's Steam Turbine.

In 1629, Branca, an Italian, invented a turbine much like a miniature water wheel, which was driven by a jet of steam from a nozzle directed against the buckets of the wheel. This is the simplest form of an impulse turbine, and is illustrated by Fig. 4.

In 1784, Wolfgang de Kempelen designed a turbine of the lawn sprinkler type, similar in principle to Hero's engine, the chief difference being the substitution of a horizontal revolving tube for the hollow sphere which Hero used. Steam, escaping from the outlets in opposite ends of the tube, caused it to revolve by reaction, just as the escaping water causes the lawn sprinkler to revolve.

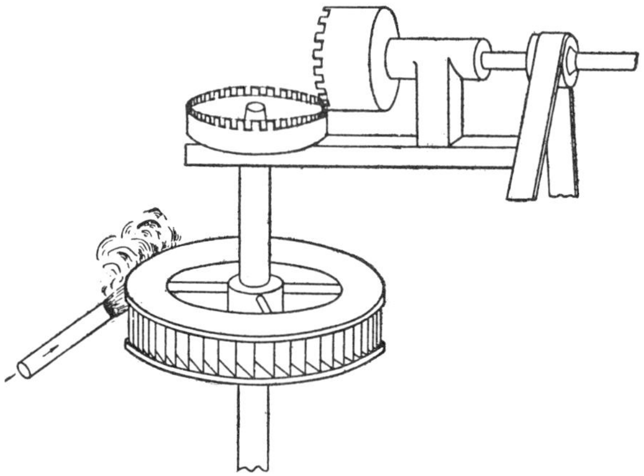

Fig. 4. Branca's Impulse Turbine.

In 1784 it is said that James Watt took out a patent on a turbine, but, as we all so well know, devoted his genius to the development of the reciprocating engine. At this time, both types of engine were in about the same crude form, and it is possible that had Watt devoted his energies to the turbine instead of to the reciprocating engine we might not have had the ordinary form of steam engine in its present

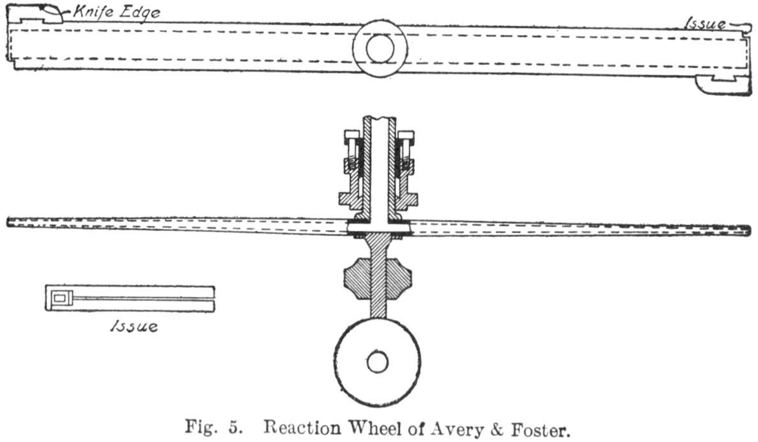

Fig. 5. Reaction Wheel of Avery & Foster.

high state of efficiency, for the turbine possesses so many advantages. This is proved by the fact that it has at last come to the front in spite of the great commercial success of the reciprocating engine.

In 1831, Avery & Foster took out the first patent granted for a turbine by the United States Patent Office. This was on the Hero lines, and was really an improvement on the Wolfgang de Kempelen turbine of 1784. This turbine appears to be the first to attain commercial success. Several were built under the Avery patent and were used to run sawmills near Syracuse, N. Y.

Steam entered a hollow shaft, Fig. 5, through a stuffing box, passed through to the hollow arms, and escaped through plain openings in opposite ends of the arms. The speed of rotation was enormous, the periphery of a 7 ft. wheel traveling at the rate of about 14 miles per minute. The wear was excessive, and this, combined with inability to get proper packing for the stuffing boxes, rather than the lack of steam economy, doubtless caused its failure, for the reciprocating engine of those days had not reached its highest state of economy. Had Avery used the present expanding nozzles instead of plain openings for his steam outlets, his steam consumption would doubtless have been less, but the speed of rotation more. Diverging nozzles were used as early as 1838, but as they were not correctly proportioned, they were a hindrance rather than a help, and the idea seems to have been given up for a time.

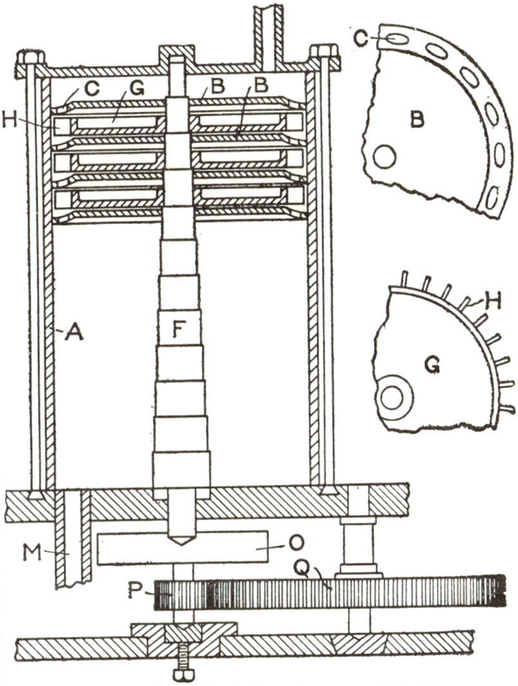

Fig. 6. Real & Pichon Compound Turbine.

As early as 1827, a compound turbine was patented by Real & Pichon, the idea being to reduce the velocity of rotation by passing the steam through successive wheels *G*, Fig. 6, separated by disks *B B* containing outlets *C* to permit the passage of the steam from one chamber to another. *H* is one of the blades, *F* the shaft, and *M* the steam exhaust. This is the principle on which the present Rateau turbine works.

The chief cause for early failure in turbine work was lack of comprehensive knowledge of the flow of steam. It was not until 1840 or thereabouts that anyone seemed to get at the real facts or appreciate the true significance of the situation. In this year Pilbrow patented a machine that was a distinct advance in the right direction, and his patent claims show that he, at least, understood some of the fundamental principles. In 1842, he attempted to reduce the speed of rotation by compounding, passing steam through

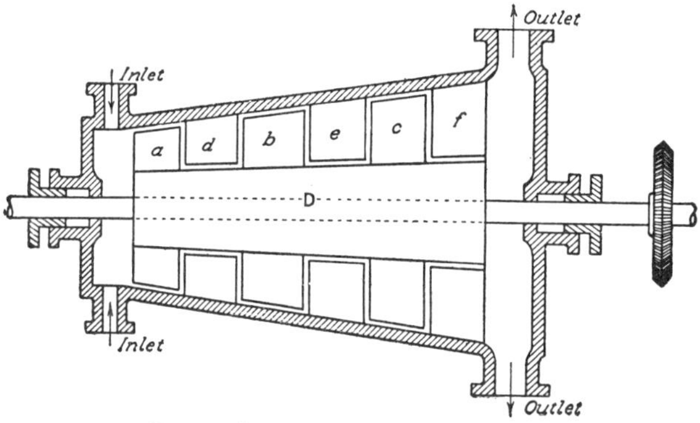

Fig. 7. Wilson's Compound Turbine.

successive wheels revolving in opposite directions. There are, of course, grave objections to such an arrangement. He later invented a turbine with several nozzles that could be successively cut out of action as the load on the turbine varied. This, in its crude form, is the fundamental idea of the arrangement of the nozzles used in the DeLaval and Curtis turbines, but Pilbrow used a converging, instead of a diverging nozzle, and his wheel was unlike others of the impulse type.

In 1840, Wilson patented the forerunner of the Parsons type of turbine. He passed steam successively through rows of running and stationary vanes, gradually expanding it until the exhaust pressure was reached. A view of Wilson's invention is shown in Fig. 7; ***a, b,*** and *c*, are vanes which are attached to and rotate with the drum *D*, while *d*, *e*, and *f* are stationary guide vanes. Steam enters at the left, passes through the turbine longitudinally, and exhausts at the right. Wilson appears to have been among the first to realize that the volume of steam increases as it expands to lower pressures, to provide for the same by increased size of passages, and, what is perhaps most important, to claim this in his patent.

In 1858, Hartman Bros. patented a turbine consisting of two revolving disks *c* and *c′* fixed to a shaft *D*, as shown in Fig. 8. Between them was a segment of stationary reversing blades *d d*. Steam entered from a nozzle *F* and was exhausted at *H*; *G* is the casing. This turbine embodied the essential element of the one-stage Curtis turbine of the present day.

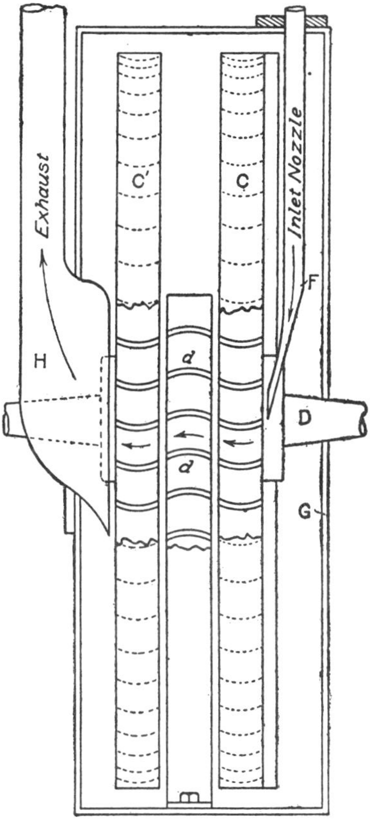

Fig. 8. Hartman's Compound Impulse Turbine.

Perrigault & Farcot, about 1870, patented a compound turbine in which the steam, as it left the buckets, passed through successive passages and again and again impinged upon the face of the same wheel. This is the principle adopted in the Riedler-Stumpf turbine and is illustrated by Fig. 9. Steam enters this turbine through the nozzle *A*, passes through the wheel buckets to the other side, and discharges into pipe *B*, which brings the steam around again to the inlet side. Here it discharges through the opening *b*, against another bucket of the same wheel, whence it is picked up by the opening *x*, in the pipe *C*, and so on, finally exhausting through the pipe *D*. Somewhat earlier, a turbine was invented that returned the steam in a similar manner, but to another set of blades on the same wheel. This idea has also been perfected by Riedler & Stumpf.

Moorehouse patented, in 1877, an improvement on the type suggested by Real & Pichon in 1827. His chief claim was an allowance for the increased volume of expanding steam in this type of turbine.

DeLaval took out a patent on a reaction turbine of the Hero type in 1883. It differed from the Avery turbine in detail, but not in principle. This turbine was extensively used for running cream separators, and was commercially successful, but was later abandoned for the present type of DeLaval motor.

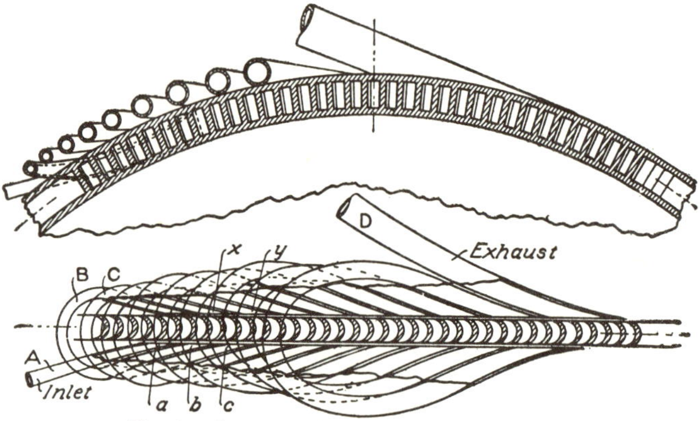

Fig. 9. Compound Turbine with One Wheel.

In 1885, Parsons took out his first turbine patent on a motor along the lines previously suggested by Wilson, and is responsible for the successful development of this type of motor. His first turbine, shown in Fig. 10, took steam in the center *A*, and exhausted at both ends through the exhaust passage *E E*, thus avoiding any end-thrust on the shaft *B*. At the same time, he patented his famous flexible bearing, now in general use. In 1888, he patented the present arrangement of grouping several rows of blades together increasing the drum diameters step by step to provide for proper expansion, at the same time patenting his balancing pistons, at present employed to relieve end-thrust.

The expanding nozzle had been patented in 1867 for use in steam injectors, but it was not until 1894 that anyone patented its use in connection with a turbine. In this year, DeLaval secured this patent and used the nozzle in connection with his turbine, for the purpose of expanding the steam and getting a high velocity of jet with increased kinetic energy.

During 1894 and 1895 there were issued a large number of patents, many of which have been successfully developed. Among them were Parsons', Rateau's, and the first patents for the use of buckets of the Pelton type.

In 1896, Curtis patented the use of an expanding nozzle in combination with a compound wheel of the type suggested by the Hart-

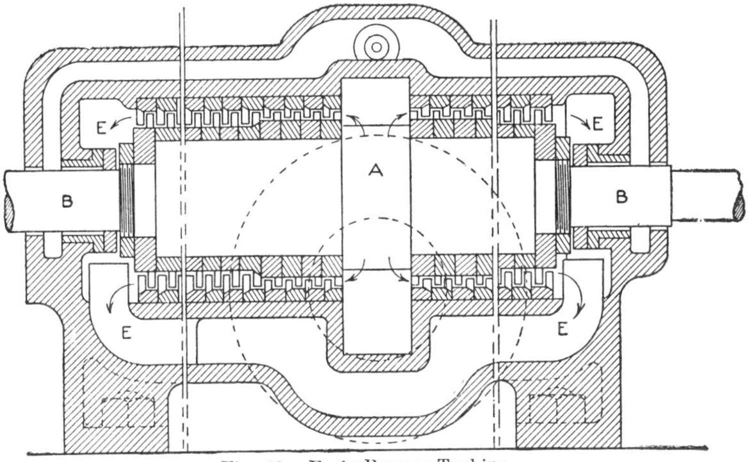

Fig. 10. Early Parsons Turbine.

man patents in 1858. Others had used both the expanding nozzle and the same type of wheel, and only two years earlier patents were taken out for a converging nozzle with a similar wheel. From a study of nozzles, it will appear that the converging nozzle could be used economically by increasing the number of stages used in the expansion, but the turbine would be larger than the Curtis, and probably less efficient.

Patents were issued in 1898 to Riedler & Stumpf, whose turbine appears to be an improvement on the Perrigault & Farcot, patented about 1870.

In 1900, the Zoelly patents were issued. This turbine in principle is similar to the Rateau, but different in construction.

The steam turbine patents issued since 1900 are altogether too numerous even to mention, but from them a number of commercial machines have been developed, and are now on the market. The principal commercial turbines will be described later.

It must not be thought that this summary is at all exhaustive, or that even all the noteworthy turbine patents have been mentioned. There are hundreds of them, and it is possible here only to mention those that are the immediate forerunners of our present commercial types. This brief summary will show that the commercial success of the turbine has been due to a more complete knowledge of the properties of steam, improved details, and the possibility of better workmanship, rather than to the development of new principles, for the distinctive fundamental ideas of all of our commercial turbines had been suggested years ago. It should be borne in mind, however, that no fundamental principle can be successfully worked out unless the minutest detail is correct, and these details may, and in the case of the turbine did, prevent the successful carrying out of the early ideas.

Fundamental Principles. The underlying principles of steam and water turbines are alike—a moving fluid impinges upon curved vanes or buckets attached to the periphery of a wheel, thus causing it to revolve. The vanes or buckets change the direction of the actuating fluid and absorb part of its energy, the fluid leaving the turbine with a comparatively low velocity. To insure reasonable economy, the fluid must impinge upon the vanes in a direction tangential to their surface at the point of impact, so as not to impart any shock and to avoid spattering. Further, the residual velocity at outlet should be as low as possible.

In either class of turbine, rotation is caused, not by the statical pressure of the actuating fluid, but by the velocity which it imparts to the rotating turbine wheels. The kinetic energy of the fluid passing through the turbine is equal to $\frac{WV^2}{2g}$, where W equals the weight of the fluid per second and V is its velocity at entrance. Evidently, the smaller W, the larger V must be to develop the same power. If the fluid leaves the turbine with the velocity V_a, then $\frac{WV_a^2}{2g}$

represents the energy not absorbed by the turbine. If V_a is small, this wasted energy will be likewise small.

Since the fundamental principles of the turbines are the same, it would seem at first sight as if steam and water turbines could be built on similar lines. But this is not so, because the difference in density and in elasticity of the two fluids requires different applications of those principles. In the steam turbine, not only must proper steps be taken to abstract the energy from the steam jet, but also to make that energy a maximum by providing for the proper expansion of the steam.

To make more clear the differences just mentioned, it should be remembered that water is an *inelastic* fluid; that is, one having a constant volume under all conditions of pressure. Therefore, in flowing through a nozzle, if the velocity at the outlet is to be greater than the velocity in the pipe, the area of the outlet must be smaller than the cross-section of the pipe. Steam, on the other hand, is an *elastic* fluid and expands rapidly as it flows through a nozzle. If the increase in volume were in exact ratio to the increase in velocity, then, for maximum efficiency, the nozzle would be parallel-sided. This happens when the pressure at discharge is about 60% of the initial pressure. But when discharging into a low pressure, the volume of the steam increases more rapidly than the velocity, and hence, if the mouth of the nozzle is to be capable of discharging the same weight of steam per second as the throat (the condition for maximum efficiency), the cross-sectional area of the nozzle must constantly increase toward the outlet.

Steam will expand as it passes through the turbine and if the passages are correctly proportioned, so that this expansion can take place only in one direction, that is, in the line of flow, the steam particles will be forced forward in a nearly uniform jet; the steam, by virtue of this expansion, will attain a very high velocity and the jet will consequently have a high kinetic energy.

Water turbines use a relatively small head and a large quantity of fluid; with the steam turbines, the quantity of fluid is small, but the head is very large. To develop large powers with any form of turbine, it is necessary that a number of wheels be used. With water turbines, each wheel acts under full head, each using a relatively small quantity of water. With steam turbines, however, it is the head that must be divided into different steps; *i. e.*, a single steam turbine can use the full quantity of fluid but, if desired to run at relatively low speeds, it can use but a portion of the total head. To develop 1,000 H. P. on a turbine shaft, with a head of 150 feet, would require approximately 4,800 pounds of fluid per second, depending somewhat upon the design of the wheel. With a head of 3,000 feet there would be required 242 pounds of fluid per second, and with a head of 234,000 feet, comparable with that of a steam turbine, the requirement would be about $3\frac{1}{2}$ pounds of fluid per second. It is thus clearly seen that the difficulty in developing large powers with the water turbine is that of providing for a sufficiently large quantity of fluid through the turbines, and with the steam turbine, that of handling the great velocities resulting from the enormous head.

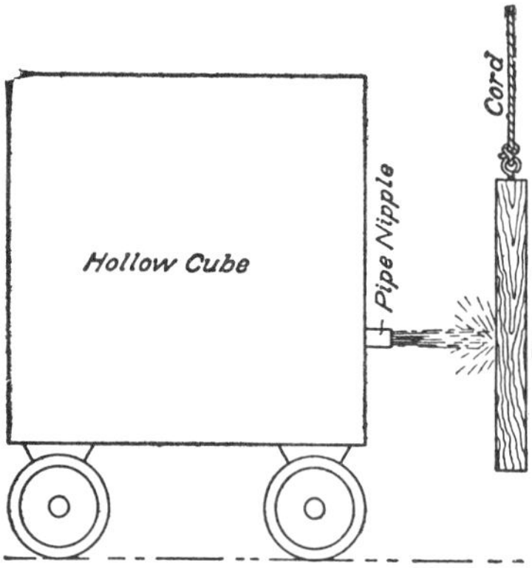

Fig. 11. Apparatus for Showing Force of a Jet.

In all steam turbines, the steam is expanded in suitable nozzles or passages. In some the expansion is all in the nozzles; in others, partly in the nozzles and partly in the vanes or blades. In some, the total expansion from boiler to exhaust takes place in one nozzle, and the energy is absorbed by a single wheel; such a turbine is called a *single-stage turbine.* In others, the expansion in one set of nozzles is only partial, and after passing through one or more wheels, the steam again passes through another set of nozzles and set of wheels, and so on, until exhaust pressure is reached. This is a *multi-stage turbine.* There may be all the way from one to forty stages, or even more, in turbines of this type where all expansion is in the stationary vanes; and in turbines where part of the expansion is in the running vanes, there may be 100 stages or more. Some turbines use nozzles for expanding the steam, and some use stationary vanes for the purpose, these vanes being so shaped as to provide suitable passage areas to permit of steam expansion. The principle is the same whether nozzles or blades are used, but blades are generally used where many stages are employed and the drop in pressure is small from stage to stage.

Before taking up the actual study of steam turbines, it will be necessary to have a clear conception of a few elementary principles of mechanics. Suppose a hollow cube to be filled with some fluid (water or steam) at a given pressure, and to have an opening in one side that can readily be closed. The arrangement is such that when the outlet is opened, the internal pressure will remain the same. If the outlet is opened, the fluid will rush out, as shown in Fig. 11; and, if the jet is supposed to strike against a board free to move, the jet will exert a force upon that board tending to swing it in the direction of the jet. This force is called an *impulse*. At the same time there will be a tendency on the part of the cube to move in the opposite direction, and the force thus developed is called a *reaction*. It may be explained in this way:

Suppose each side of the cube to be one foot square, the area of the opening, one square inch, and the internal pressure, 100 pounds per square inch. There will be 144 × 100 pounds pressure on each side of the cube with the outlet closed, but when the one-inch outlet is opened, the total pressure on the side containing the outlet will be reduced by the pressure of 100 pounds on the opening itself. This will leave an unbalanced force of 100 pounds acting in the opposite direction, which is the origin of the reactive force. This explanation is not strictly correct, but serves to give an idea of these two forces, impulse and reaction.

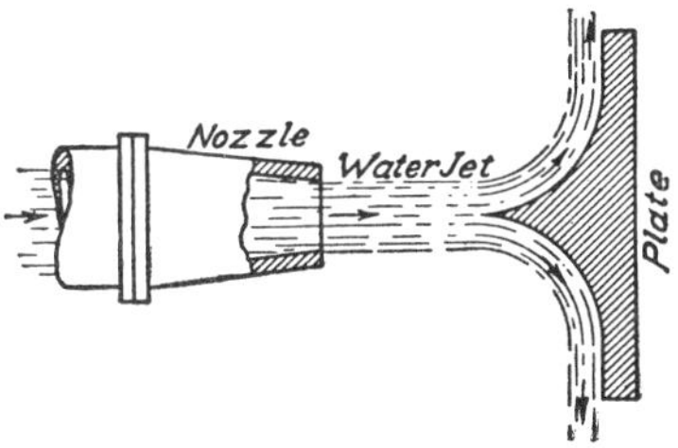

Fig. 12. Jet Deflected through 90°.

Hero's turbine was a reaction turbine pure and simple, Branca's, an impulse turbine; but what is called a reaction turbine at the present time is not a simple reaction turbine in any sense, but one running under the combined influence of reaction and impulse. Likewise, the so-called impulse turbine is not a pure impulse turbine, but acts under the combined influence of impulse and reaction. There is no pure reaction turbine now on the market. The so-called impulse turbine being rather simpler of explanation, for the present only this type will be considered in the following explanations. How these principles apply to the so-called reaction turbine will be explained later.

Suppose a stream of water from a nozzle to impinge upon the plate shown in Fig. 12, and so made that the jet is divided, and without shock departs in a direction tangential to the plate and at 90° to the line of impact. If the velocity of impact of the jet is V feet per second, its velocity in the same direction after striking the plate will be zero, and therefore, a definite force will be exerted on that plate, equal to the force necessary to impart a velocity of V feet in one second to the mass of water in the jet. The acceleration, therefore, will be V feet per second, and since force is measured by mass times acceleration, this force, acting on the plate, will be $F = MV$. If the plate is allowed to move in the direction of the jet with a velocity V_1, the relative velocity of the plate with reference to the jet will be $V - V_1$, and the corresponding force acting on the plate will be $F = M(V - V_1)$. Since work is measured by the product of force and distance, the force acting through the space V_1 in one second, will do the work $W = FV_1 = M(V - V_1)V_1$ foot pounds.

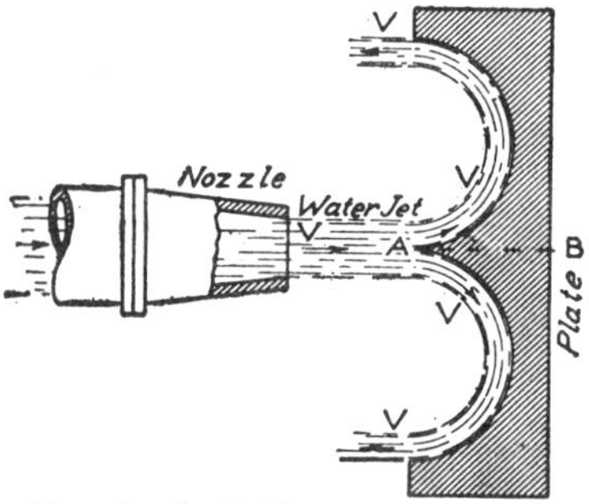

Fig. 13. Jet Deflected through 180°.

Now if the plate were shaped as shown in Fig. 13, so that the direction of the jet were completely reversed, that is, turned through 180°, there would be an additional pressure on the plate, due to the reaction of the jet leaving it. This, neglecting friction, would be equal to the original impulse, thus making the total force on the plate 2 F instead of F. It is quite evident that if the force is twice as great, the work must also be double, and the above expression for the work done becomes

$$W = 2FV_1 = M \times 2(V - V_1)V_1$$

For this reason, turbine vanes are made so as to reverse the direction of the jet as completely as possible. Complete reversal is not practicable because some clearance must be allowed for the deflected jet to escape. This is especially true in the usual case in practice, where the jet impinges upon the vanes from the side. Here, the angle has to be such that the revolving wheel will clear both nozzle and deflected jet.

If the bucket shown in Fig. 13 were held stationary, the force exerted by the jet would evidently be a maximum and equal to $2MV$; but the velocity of the bucket being zero, the work, equal to the force multiplied by the space, would also be zero. If, on the other hand, the velocity of the bucket were equal to the velocity of the jet, the push would be zero, and the work again zero. Somewhere between these limits, there must evidently be a velocity which will produce maximum results.

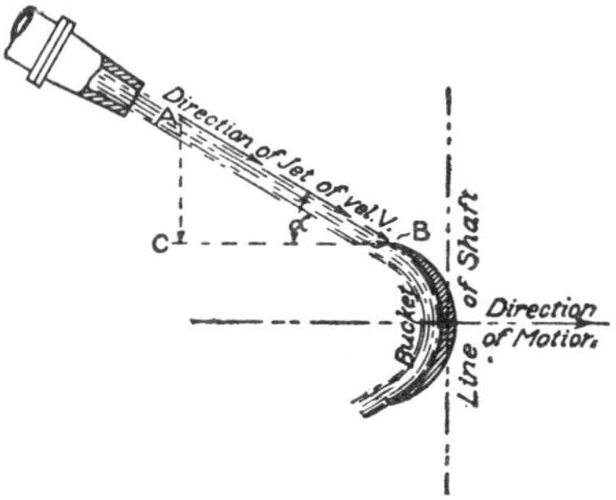

Fig. 14. Jet Impinging upon Curved Vane at an Angle with Plane of Rotation.

Suppose now, that $V_1 = \frac{V}{2}$; then

$$W = M \times 2\left(V - \frac{V}{2}\right)\frac{V}{2} = M\frac{V^2}{2},$$

but

$\frac{MV^2}{2}$ = the kinetic energy of the jet, as it issues from the nozzle. Therefore, if the speed of the bucket is one-half the velocity of the jet, we have an efficiency of 100%, neglecting losses, and this is, of course, the best obtainable. Therefore, the greatest efficiency is obtained when the speed of the bucket is half the jet velocity, provided the jet impinges upon the bucket in a direction parallel to the line of movement of the bucket. For other angles, the speed for maximum efficiency would be somewhat less.

If a jet with the velocity V strikes the bucket at an angle α, as shown in Fig. 14, its velocity AB could be resolved into two components—one CB at right angles to the shaft, and one CA parallel to the shaft. The one at right angles to the shaft, commonly known as the *velocity of whirl*, would produce a rotative impulse equal to $V \cos \alpha$, and V_1, the velocity of the vane necessary for maximum efficiency, would be half this, or $V_1 = \frac{1}{2} V \cos \alpha$, provided the angle with which the jet leaves the blade is equal to the angle of impingement. The component $A\ C$, parallel to the shaft, would have no tendency to cause rotation, but would produce an end thrust on the shaft. This component is called the *velocity of flow.*

Suppose a jet to impinge upon a curved vane at the angle shown in *a* Fig. 15. If the jet strikes this vane tangentially, without shock, the vane remaining stationary, the relative positions of the jet before and after impact will be as shown. Now if the vane is allowed to move with the velocity V_1, the relative positions of the vane and the nozzle will change, and the jet will no longer glide smoothly onto the vane, but will strike the edges, and spatter. To maintain the correct relative positions, the nozzle must either be allowed to follow the vane, or its position must be changed so that the direction and velocity of the jet will be such that it may be resolved into two components, one parallel with the direction of motion of the vane, and the other tangent to the vane. The absolute direction of the jet must be along the line $A\ B$, (b, Fig. 15), but its direction relative to the moving vane will be along the line $A\ C$, and if $A\ B$ is drawn to a scale representing the actual velocity of the jet, and $C\ B$ laid off to the same scale to represent the velocity V_1 of the vane, then $A\ C$ will represent in magnitude and direction the relative velocity of the jet and the vane, which will be identical with the absolute velocity in the first case where the vane is stationary. Neglecting friction, the jet will leave the vane with the same relative velocity. Draw $E\ F = A\ C$ and $E\ G = C\ B =$ the velocity V_1. Then $E\ H$, which we shall call V_3, will represent in magnitude and direction the absolute velocity with which the jet leaves the vane.

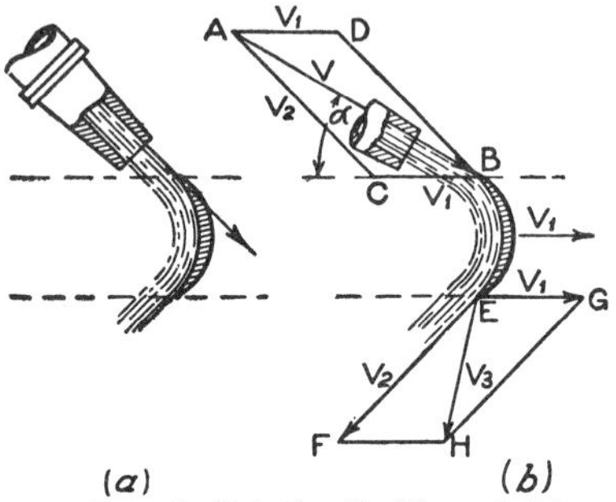

Fig. 15. Relative Positions of Jet and Vane.

The energy in the jet before impact was $\frac{W\ V^2}{2g}$,

after leaving the vane, $\frac{W\ V_3^{\ 2}}{2g}$.

The energy absorbed was then $\frac{W\ V^2}{2g} - \frac{W V_3^{\ 2}}{2g} = \frac{W}{2g}(V^2 - V_3^{\ 2})$.

For the best efficiency, V_3 should be small, but can never be zero unless the jet angle a is zero, and the direction of the jet is reversed through 180°, an impracticable condition.

Nozzles. Steam does not cause rotation in the turbine because of its statical pressure, but, as already stated, because of its velocity, a difference in pressure acting indirectly, by imparting velocity to the steam. It is evident, then, that in this class of motor, steam velocities are all-important. The steam possesses energy by virtue of the heat which it contains, but to make this energy available in the turbine, it must be transformed into kinetic energy by the production of a high jet velocity. The correct shaping of the nozzle is the all-important factor in acquiring the requisite steam velocity, as will appear from the following considerations:

It has been well established by experiment that steam at high pressure flowing into a space at lower pressure, through a nozzle with parallel sides, cannot attain a velocity exceeding 1,450 to 1,500 feet per second, no matter how high the initial pressure nor how low the pressure into which the steam discharges. This limiting velocity is due to the fact that at the throat of any nozzle there occurs a drop in the pressure of the steam to about 58% of the initial pressure, and, if the nozzle be a cylindrical one, this drop will remain practically constant throughout the length of the nozzle. The velocity acquired by virtue of this difference in pressure will therefore be about the same, whether the absolute pressure into which the steam is discharged is 58% of the initial pressure or much less. In the latter case, the throat pressure cannot change until the outlet is reached, when the pressure drops suddenly to the pressure of the space into which the steam is discharging, and the steam immediately expands in all directions, thus dissipating its energy. The only case in which maximum efficiency is developed with orifices and short passages with parallel sides is when the low pressure is greater than 58% of the high pressure.

This limiting value for orifices or parallel-sided nozzles, and the consequent limit of steam velocity; makes it impossible to develop the greatest energy of the steam when expanding to low pressures except through a nozzle with flaring sides, in which the outlet is greater than the inlet. In such a nozzle the steam expansion occurs gradually in its flow, and is constrained to take place only in the direction of the flow. In this way, the velocity of the steam particles is increased as it proceeds along its nozzle until a tremendous speed has been developed, which will produce about 95% of the available energy. Furthermore, steam may be expanded effectively within the confines of such a nozzle *from any high pressure to any lower pressure,* provided the increase of areas of cross-section of the nozzle is proportional to the increase of specific volumes of the steam.

In other words, with a cylindrical nozzle, a limiting steam velocity of 1,450 to 1,500 feet per second is possible, no matter whether the initial pressure of the steam be 70 pounds or 200 pounds, or whether the pressure into which the steam is finally expanded be 58% of the initial or a 28″ vacuum. Of course, as the weight of the steam per cubic foot varies with the pressure, a greater *weight* of steam will be discharged per second at higher pressure, resulting in a somewhat greater kinetic energy in the steam jet. On the other hand, if the nozzle has flaring sides, the steam at, say, 150 pounds gauge pressure, will have the same pressure at the throat, that is, 58% of the initial pressure, but will acquire a rapidly increasing velocity from throat to outlet, and, with a 28″ vacuum ahead of it, will leave the nozzle with a velocity of 4,000 feet per second, assuming no friction in the nozzle. Fig. 16 shows a properly designed nozzle for expanding steam from 150 pounds boiler pressure to 28″ vacuum.

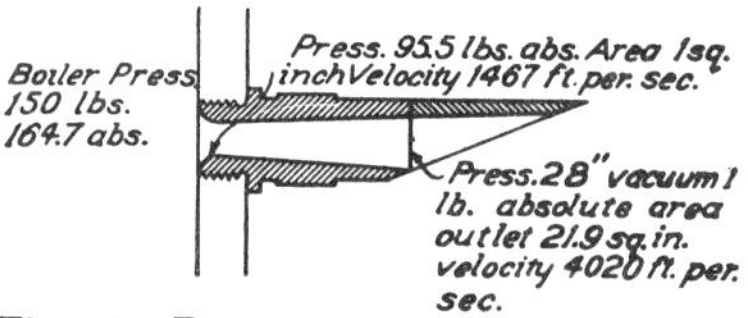

Fig. 16. Properly Designed Expanding Nozzle.

Compounding. It has been explained that if steam is expanded in a suitable diverging nozzle, nearly all the heat energy becomes available as kinetic energy, and that this steam, when flowing from a boiler pressure of 150 pounds to a vacuum of 28″, may attain a velocity of approximately 4,000 feet per second. If the linear velocity of the buckets were to be approximately one-half the velocity of the jet, there would be grave danger that the wheel would burst from centrifugal force. A peripheral speed of 1,200 feet per second is the limit in practice. 2,000 feet per second would mean about 12,750 revolutions per minute in a wheel 3 feet in diameter. This latter velocity would mean great delicacy in balancing and difficulty in providing suitable bearings, even if material could be found to withstand the strain. A wheel 15 feet in diameter would have to revolve 2,500 revolutions per minute if the above mentioned peripheral velocity were to be obtained, and again the impossibility of construction is evident. Some means must therefore be employed to reduce the speed to manageable rates without unduly increasing the size of wheel. This may be done in a single-stage turbine by means of gearing, but here, if 1,200 feet is to be the maximum permissible peripheral velocity, and 2,000 is the theoretical velocity, there will be a loss of efficiency. As a matter of fact, the steam jet does not strike the wheel in a line at right angles to the shaft; consequently the velocity of whirl, as already seen, is $V\cos\alpha$, and with friction allowance, this is somewhat reduced, but, even with a bucket velocity of 1,200 feet per second, the revolutions will usually be too high.

Turbine speeds may be satisfactorily reduced without the use of gearing, by what is called *compounding; i. e.*, by dividing expansion into separate stages, called *pressure* compounding; by passing the steam over several wheels with guide vanes between to redirect the steam upon the vanes in the next wheel, called *velocity* compounding; or by a combination of these two methods.

Suppose we start with steam at 150 pounds gauge pressure and expand it to 28 in. vacuum, not in one expanding nozzle but in several stages, so that the expansion in each would be to only about 60% of the next higher pressure, in which case, diverging or expanding nozzles would not be needed. The velocity of flow of the steam would be somewhat less than 1,450 feet per second at pressures above the atmosphere, and would decrease slightly as the pressure lowered; the lowest velocity, when discharging into a vacuum of 28 in., would be about 1,250 feet per second, but, by letting the drop in pressure be somewhat less than 60% in the higher stages, the velocity of flow could be made approximately 1,250 feet per second throughout. This is, of course, neglecting all losses. We could then have a steam speed of about 1,250 feet per second to deal with, instead of 4,000; the peripheral speed of the buckets would be $\frac{1{,}250\cos\alpha}{2} = 625\cos\alpha$, or, when α, is small, about 600 feet per second. For a wheel 5 feet in diameter, this would mean about 2,300 revolutions per minute, and the conditions arising from such a speed are much more easily taken care of. This reduction in speed could be accomplished in about ten stages. To reduce the speed to half, or 300 feet per second, would require four times as many stages because the number of stages would be equal to the square of the ratio of reduction of the steam velocity. Thus, the reduction from 4,000 feet per second to 1,250 feet per second will require $\left(\frac{4{,}000}{1{,}250}\right)^2 = \overline{3.2}^2 = 10$, approximately. To reduce the speed of the buckets from 600 to 300 feet per second, would evidently require four times as many, or 40 stages.

Suppose, now, the compounding were all in velocity stages, and that the expansion occurred in one nozzle. The velocity of steam would be nearly 4,000 feet per second, but it would have to pass through three sets of revolving wheels to bring the relative speed at each wheel to approximately 1,300 feet per second, and thus get the same speed of revolution as in the previous case.

This method of compounding gives a very compact form of turbine and one that has many mechanical advantages; but the wheels have to revolve in a bath of steam which makes the friction excessive, and the efficiency correspondingly lower. This was the idea of the original Curtis patent, but was soon abandoned for the combined pressure and velocity turbine which is to-day the principal feature of the Curtis design. In this combined method, there are two or more pressure stages, and in present practice, not over two velocity wheels and one set of guide vanes to each stage.* The older Curtis had even three or four revolving wheels per stage with a corresponding number of sets of guide vanes.

Compounding has been tried by the use of counter-running wheels, but with little success. If the guide vanes were on wheels free to turn, they would run in a direction opposite to the others, and if the relative peripheral velocity of the two were, say, 1,200 feet per second, it would mean that for each wheel the absolute velocity would be half this, or 600 feet per second. The difficulties of building and operating such a machine are considerable.

Types of Turbines. There are two main groups into which steam turbines are usually divided, one known as the *impulse,* and the other as the *reaction* type. It has become the general practice

*This does not apply to marine practice, the peculiar conditions of which warrant the use of a larger number of velocity wheels per stage. Small Curtis turbines and some special machines have three velocity stages.

to classify turbines under one or the other of these two general heads; but, as a matter of fact, every commercial turbine of the present time really develops its power under the combined influence of action and reaction. Yet there is a distinct difference between the expansion of steam in these two types, as, for instance, in the DeLaval and in the Parsons turbine. The use of the terms *impulse* and *reaction* in reference to turbines is undoubtedly unfortunate, but since their use has become practically universal, it is necessary to understand the significance of their application.

In the so-called impulse turbine, the steam, expanding in a nozzle or other suitable passage, thus attains a high velocity, and impinges upon the vanes of a rotating wheel. The steam, in passing through the wheel, gives up a part of its kinetic energy to the revolving vanes, and leaves the wheel at a lower velocity, but at the same pressure at which it left the nozzle. In the so-called reaction type, the steam enters the turbine at boiler pressure, passes through guide passages onto the vanes of a rotating wheel, and little by little expands as it passes through these vanes to subsequent guide passages and other vanes, the pressure gradually becoming lower; the velocity which is gained by the expansion in the guide passages and revolving vanes is practically all imparted to the rotating drum. The pressure is less on the one side of the vane than on the other, while, in the impulse type, the pressure is the same on both sides of the vane.

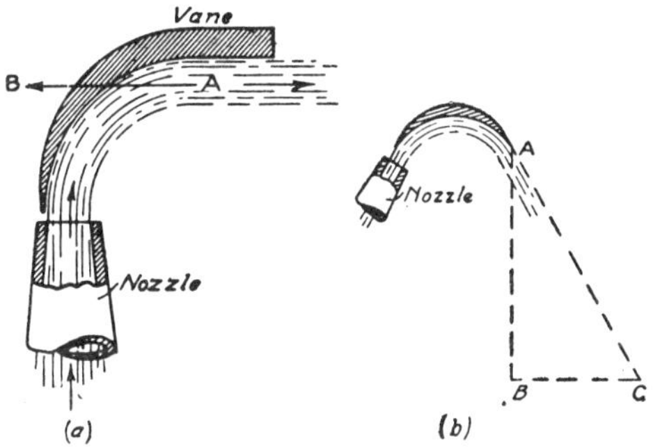

Fig. 17 Two Arrangement of Jet and Vane. (*a*) Pure Action. (*b*) Action and Reaction.

The engines of Hero, Wolfgang De Kempelen, and Avery were all purely reaction types, but the Parsons acts by impulse as well as reaction, and the Curtis and DeLaval, by reaction as well as by impulse. To make this clear, consider Fig. 17 (a), which shows a vane and jet. The vane is so shaped that the jet leaves it at right angles to the direction of impact. Here is a case of pure action, so far as any force tending to move the vane in a direction parallel to the direction of the jet is concerned. There is, to be sure, a reaction of the jet, but this reacting force is along a line $A\ B$ at right angles to the desired line of motion, and if the vane shown in this figure were attached to the periphery of a wheel free to revolve, this force of reaction would cause only an end-thrust on the shaft, in no way augmenting the force of rotation.

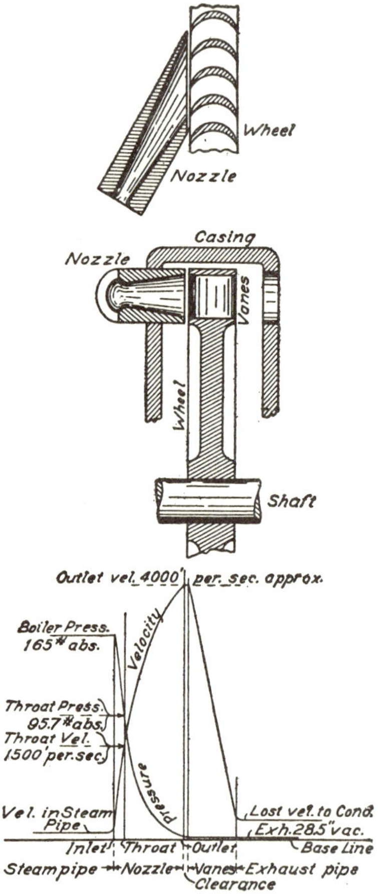

Fig. 18. Typical Features of Single-Stage Impulse Turbine, with Relations of Steam Pressures and Velocities.

As previously shown on Page 16, to obtain the best efficiency the jet must be deflected through an angle of 180°. If the jet leaves the wheel at any less angle than 90°, for instance, angle $B\ A\ C$ in Fig. 17 (b), there is a reactive force along the line AC, which can be resolved into two components —one, $A\ B$, tending to cause rotation, the other, $B\ C$, causing an end thrust. A turbine thus constructed, although called an impulse turbine, evidently derives an impelling force from this reaction. A pure impulse turbine, permitting no reaction of the jet, would have a theoretical maximum efficiency of only 50%. When the jet is turned through an angle of 180°, the reaction becomes equal to the impulse. The reaction is equal to the impulse in any case, when the angle at which the jet impinges upon the vane is equal to the angle of deflection measured from a plane through the center of the rotating wheel at right angles to the shaft. In this type of turbine, all the expansion takes place in the nozzles or guide passages, none at all in the revolving vanes.

In the so-called reaction turbines, the expansion takes place in the revolving vanes as well as in the guide passages, and the vanes and guides are placed so as to give a constantly increasing area of passage to allow for the increasing volume of the steam as it expands. As the steam expands in the guide passages, it acquires velocity and impinges upon the running vanes, thus giving a decided impulse to them, and as it again expands in the running vanes, the reaction produces a further impelling force.

The distinguishing feature, then, between these two distinct types of turbine is not to be found in the impulse or reaction of the steam at all, for both types, as we have seen, act by virtue of both forces; but the distinction lies in whether the expansion of the steam takes place fully in a set of nozzles or guide passages with no expansion in the moving vanes, or whether the steam expansion takes place partly in the nozzles and partly in the revolving vanes. A turbine might be so arranged that the expansion would take place entirely in the moving vanes, the guide passages acting merely to change the direction of the steam, but as yet no commercial turbine has been built on these lines.

There are several distinct subdivisions of the two main types of turbine. The simplest form is undoubtedly of the DeLaval type, which consists of several diverging nozzles, expanding the steam from boiler pressure to exhaust pressure, and directing the steam jets onto the vanes of a single wheel. We have seen that the enormous velocity of 4,000 feet per second will be attained in expanding from 150 pounds boiler pressure to 28½ inches vacuum. The speed of revolution must be very high and, although the velocity is greatly lowered as the steam passes through the wheel, it will leave the wheel with a considerable residual velocity which represents, of course, so much lost energy. Fig. 18 illustrates the typical features of this style of machine, the curves showing the relation of its steam velocities and pressures. It will be noticed that the steam pressure is a maximum, and equal to boiler pressure, at the inlet to the nozzle, and will reach the condenser or exhaust pressure at the nozzle outlet, as it impinges upon the vanes of the wheel. Clearance, in this type of turbine, is of small consequence, for the wheel revolves in steam of a uniform pressure, and there can, therefore, be no leakage of steam without work being done. As there is but one wheel revolving in the bath of steam, the friction would not be very great, were it not that the friction increases very rapidly with the speed, and in this single-wheel type, the speed of the steam is very high. The chief loss will be due to the relatively high velocity of the exhaust steam, and to the friction of the bearings on account of the high rotative speed. To reduce these speeds of rotation to man-

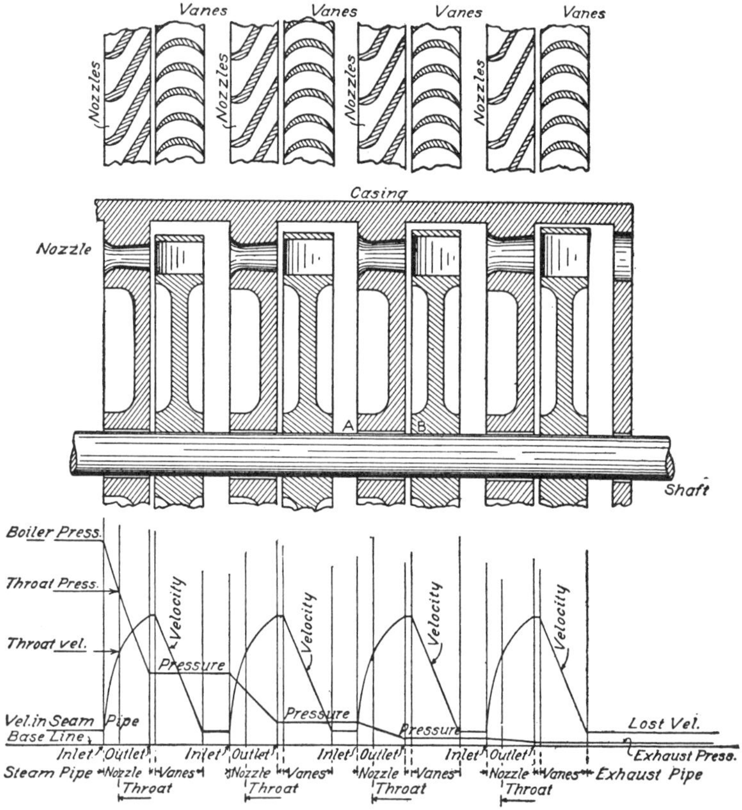

Fig. 19. **Features of Multi-Stage Impulse Type, Showing Relations of Steam Pressures and Velocities.**

ageable rates, gearing must be used, causing a further frictional loss, or the diameter of the wheel must be abnormally great.

The velocity of the steam at the entrance to the nozzle is that due merely to its flow through the pipe. At the throat of the nozzle, the velocity, as we have previously seen, will be something under 1,500 feet per second, and at the mouth of the nozzle, if it is properly designed, the velocity will approximate 4,000 feet per second, assuming a boiler pressure of 165 pounds absolute and $28\frac{1}{2}$ inches of vacuum. This high velocity will not be maintained, however, as the steam passes through the revolving vanes, but, at the condenser, will have dropped to a value depending upon the amount of energy absorbed from the steam during its passage through the vanes of the wheel.

If wheels were arranged in successive chambers, so that the steam could be expanded in several steps instead of in one, we should have the essential elements of the Rateau type of turbine. Fig. 19 shows diagramatically the essential features of this type of turbine, and the relation of velocities and pressures, as before. Each wheel rotates in an independent chamber separated from the next by a diaphragm provided with suitable expanding passages, so that the steam, in passing from the first chamber to the second, will be under conditions similar to those obtaining when passing from the boiler into the first chamber, and again may attain a maximum velocity.

In a four-stage turbine of this sort, the pressure, as shown in the curve in Fig. 19, should be a maximum (boiler pressure) at the inlet to the first nozzle. At the throat of the nozzle, it should be approximately 58% of the initial pressure. During its passage through the first chamber, the steam pressure would be constant, and it would again drop in a similar manner, in passing through the nozzles between the first and the second chamber, the velocity rising with each drop in pressure. With a four-stage turbine, the drop in pressure would be such that one-fourth of the total available heat units would be available in each chamber. The drop in pressure from chamber to chamber would therefore not be uniform, for a given pressure change represents more heat units in the lower than in the higher ranges of pressure. The velocity at the inlet to the first nozzle would again be merely the velocity of flow through the steam pipe; at the throat of the nozzle, approximately 1,500 feet per second, and at the outlet to the nozzle, where the steam impinges upon the vanes of the first wheel, approximately 2,000 feet per second. This velocity will drop as the steam passes through the wheel, rising again on its passage through the next nozzle, dropping again in the next wheel, and so on, the residual velocity as the steam leaves the last wheel being probably less than in the previous case.

Turbines are built on this principle by a number of manufacturers, the Rateau being the best known of this type. This particular turbine has usually a large number of chambers, frequently 30 to 40, and the drop in pressure from chamber to chamber is consequently very small, so small in fact that expanding nozzles are not necessary. This type of turbine is subject to leakage at *A B*, Fig. 19, where the shaft passes through the stationary diaphragm and requires special packing. This packing becomes evidently inaccessible in a multi-stage turbine.

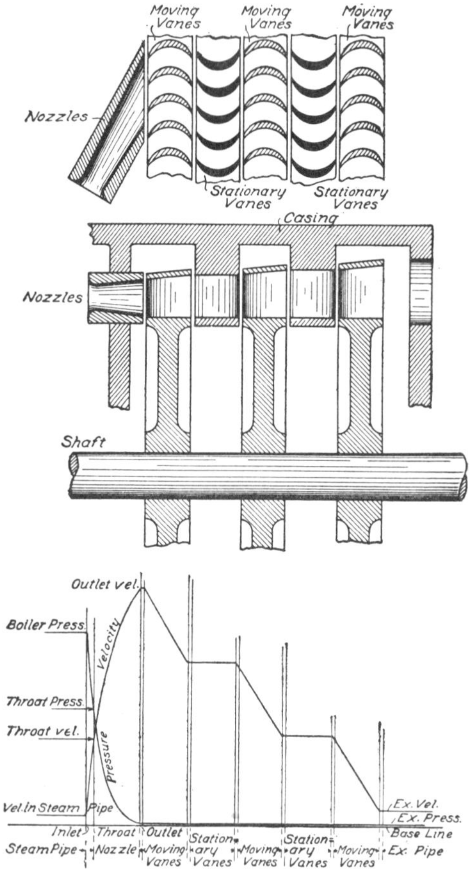

Fig. 20. Features of Impulse Turbine Compounded by Velocity Steps Only.

A simple method of compounding, but one not likely to produce as economical results, is that shown diagramatically in Fig. 20, its variations of pressure and volume being shown in the curve. In this turbine, steam is expanded in a properly designed diverging nozzle, from boiler pressure to exhaust pressure, and impinges successively upon the vanes of rotating wheels. Between these wheels are stationary guide vanes curved in the opposite direction, so that, as the steam leaves the first set of vanes, it is redirected by these guides upon the next set, and so on. The boiler pressure is exactly similar to the boiler pressure shown in Fig. 18; the velocity of the steam as it leaves the nozzle is also the same. This velocity drops somewhat as the steam passes through the first set of running vanes, remains constant as it passes through the first set of guide vanes, again drops in the next set of running vanes, and again becomes constant in the guide vanes, and so on; the velocity of the steam jet is gradually lessened as it passes through wheel after wheel. The drop in velocity in the steam in its passage through any one set of vanes will, neglecting losses, be approximately equal to the total velocity divided by the number of sets of running wheels. In this type of turbine, since the velocity of the steam is gradually decreased, it is evident that if the same quantity of steam is to flow through successive wheels in the same interval of time, the passages must gradually increase in size. The velocity remaining constant in the guide vanes, they may provide passages of uniform section, as shown in Fig. 20, each set of passages, however, being larger than the preceding set.

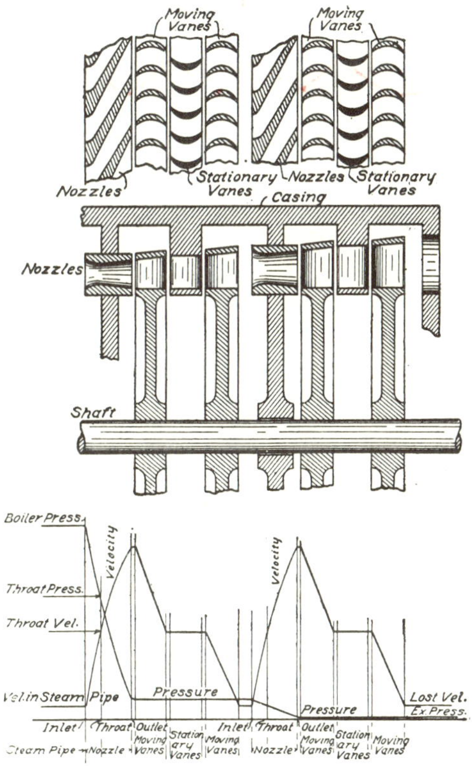

Fig. 21. Features of Turbine Compounding by Pressure Stages and Velocity Steps.

The principle just described was the original idea claimed in the early Curtis patent, but was subsequently given up for the improved arrangement shown in Fig. 21. This arrangement differs from the other, in that, instead of fully expanding the steam in one nozzle or set of nozzles from boiler to exhaust pressure, the expansion is divided into two or more stages. This turbine contains chambers, just as the Rateau type does, the difference being that in the Curtis, each chamber contains two sets of running wheels and one set of guide vanes, while the Rateau chamber contains only one wheel and no guide vanes. Turbines of the Curtis type have from two to seven pressure stages, but at the present time, no more than two sets of running vanes are used in each chamber*, although formerly, more sets of running vanes were used. The relation of pressures to velocities shown in Fig. 21 will be evident from the previous explanations.

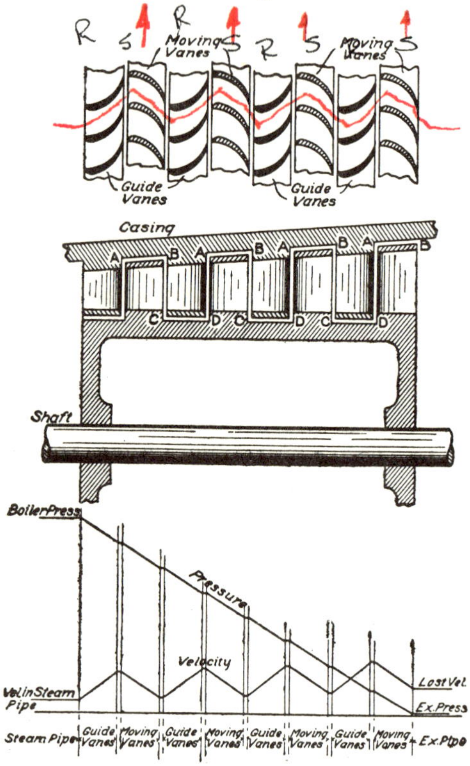

Fig. 22. Features of Reaction Type.

In the reaction turbines, of which Parsons' is the best known example, the steam, as already stated, gradually expands in passing from boiler to condenser pressure. The velocity rises in the first set of stationary vanes, and drops as the steam does work in the first set of running vanes. The velocity rises again in the next set of stationary vanes, drops in the moving vanes, and so on. Fig. 22 shows the essential features of this turbine and the relation of pressures and volumes. The stationary guide vanes act just like small nozzles, and allow the steam to expand and acquire velocity. The moving vanes also allow the steam to expand, and the reaction of this expansion gives an added impulse to the rotating wheel.

It is not intended that the foregoing shall be a description of any turbine, but merely a description of the distinct and elementary features of the action of steam in various types of turbine.

*See foot-note, Page 22.

In the one-stage, compound-velocity turbine, the steam leaves the nozzle at exhaust pressure with a high velocity. If the expansion has been complete, as intended, the pressure remains constant as the steam passes through the turbine, and there is no tendency to leakage. The clearances between the blade tips and the casing can be made as large as convenient, for it requires a difference of pressure to cause steam leakage. If running on vacuum, there would be a tendency for air to leak in around the shaft, and consequently this would need to be well packed.

Here would seem to be a happy solution to the problem of steam leakage, at the same time producing a most compact form of turbine; but, unfortunately, a considerable loss is brought about by steam flowing past the surfaces of both moving and guide vanes and by the large amount of friction, due to the rotation of the many wheels through the steam which fills the turbine. A further serious disadvantage is that an equal amount of work cannot be done in each set of vanes if the entrance and exit angles of the vanes are made equal, as is usually the case. For example, suppose a 4,000 foot steam velocity to be reduced in four wheels, each wheel absorbing 1,000 ft. per sec. Then, if V_1, V_2, V_3, and V_4 represent the respective velocities at the entrance of each wheel, the available energy is

for the first wheel, $\frac{WV_1^2}{2g} - \frac{WV_2^2}{2g} = \frac{W}{2g} \times 7{,}000{,}000;$

for the second, $\frac{WV_2^2}{2g} - \frac{WV_3^2}{2g} = \frac{W}{2g} \times 5{,}000{,}000;$

for the third, $\frac{W}{2g} \times 3{,}000{,}000;$

and for the fourth, $\frac{W}{2g} \times 1{,}000{,}000.$

This difficulty will be remedied by increasing the number of pressure stages, and decreasing the number of wheels in each stage to a minimum. With a large number of velocity compound wheels, the work done by the last wheel would be so small that the frictional losses would be too large to make it at all economical. For example, with six wheels, the last wheel would develop only 9% of the power developed in the first. In turbines of this type, by a suitable design of the nozzles and entrance and exit angles of the vanes, the same amount of steam energy may be abstracted in each pressure stage.

The leakage in this type would be relatively small, only what would pass from stage to stage. This would be comparatively small, because the steam could escape only through the opening where the shaft passes through the diaphragm (*A B* Fig. 19) that separates the two chambers, and with small clearances this could not be large. With a large number of stages, as in the Rateau turbine, leakage in the high pressure end is not all lost, for it has an opportunity to work in the lower stages.

In the reaction turbine, leakage of steam is a most important factor. As the pressure on the two sides of the vane is different, there is a tendency for the steam to escape between the tips of the vanes and the outer casing *A B*, Fig. 22, also between the ends of the guide blades and the rotor *C D*, Fig. 22. As the rotors are of large diameter, a large area is offered for leakage, unless the clearances are kept very small. Here, the steam leaking from the higher pressures, will of course do work on the lower pressure, but at a less efficiency, just as in the Rateau type. The successful turbine of this type requires great nicety of workmanship in order that the clearances may be adjusted to a minimum.

Low=Pressure Turbines. The greatest drawback to improvement in any existing engine plant, or, in fact, in any mechanical installation, has always been the fact that the equipment already installed must be discarded, often thrown into the scrap pile, while still in fairly good condition and capable of doing a considerable amount of work. In the early installation of steam turbines, this was often done, and in order to increase the capacity of the central station, good reciprocating engines were often thrown out and turbines put in their places. It was, however, soon discovered that this, in many cases, was unnecessary, and that the desired increase in power could be had by simply using low-pressure turbines in connection with the existing reciprocating engines. The low-pressure turbine takes the steam exhausted by the engine, slightly above the atmospheric pressure, and expands it to a lower vacuum than could be economically done in the engine.

While the reciprocating engine is highly efficient for utilizing the available energy of steam between boiler and atmospheric pressure, it is relatively inefficient for utilizing the energy of steam in the lower ranges of pressure, especially at pressures below 20 in. vacuum. The steam turbine, on the other hand, utilizes the available energy of steam in the lower more effectively than in the higher ranges of pressure. Since there is about as much available energy in steam below the atmospheric line as there is in steam above it, there is every reason to believe that this combination of engine and turbine will be a most efficient one. In order that the possibilities and limitations may be fully stated, however, it will be necessary to investigate some of the characteristics of steam expansion.

A single cylinder engine with cut-off at, say, one-third stroke, will expand the steam to three times its initial volume, and if it takes steam at 150 pounds gauge pressure, the volume of each pound of that steam before expansion will be approximately 2.75 cubic feet. Now, if this is expanded to three times its initial volume, every pound of steam entering the cylinder will, at exhaust, occupy $3 \times 2.75 = 8.25$ cubic feet. If the expansion has been adiabatic, that is, without the gain or loss of heat, this pound of steam will occupy 8.25 cubic feet of space when the pressure has reached 32 pounds by the gauge, and, under the above conditions, an engine would release at this pressure—a manifest waste.

With one-fifth cut-off and five expansions, the final volume of one pound would be $5 \times 2.75 = 13.75$ cubic feet, and this volume would be reached at about 11.7 pounds gauge pressure. Fig. 23 will illustrate this. The line *b c d e* is a curve representing the relation of pressures and volumes of steam, as it expands adiabatically from 150 pounds gauge pressure to the atmosphere and beyond the atmosphere into partial vacuum. The total available work in the steam above atmospheric pressure, would be represented by the area of the diagram *a b e h*. The greatest possible work that could be done in the cylinder, cutting off at one-third stroke and exhausting at atmospheric pressure, would be the area *a b c g h*, which shows that a considerable amount of the energy is lost. Even cutting off at one-fifth stroke, the work represented by the area *d e f* is lost. To carry the expansion of steam in a single-cylinder engine even to one pound above the atmosphere, the boiler pressure must be greatly reduced, or the amount of expansion increased materially. If this engine were made condensing, *m k* would represent the back-pressure line, and while the total available energy would be increased by the area *h e m k*, the work in the cylinder at one-fifth cut-off would be increased only by the area *h f j k*, a very small part of the whole. In such case, the gain would probably not pay for the cost of maintaining the vacuum.

In a compound two-cylinder engine taking steam at 150 pounds gauge, the ratio of high- to low-pressure cylinder volumes would be not over 1 to 5, and with cut-off on the high-pressure cylinder at one-third stroke, there would be room for not over 15 expansions; that is, the volume of steam at the end of the low-pressure stroke would be not over 15 times the volume of the steam admitted. Now, if one pound of steam at 150 pounds gauge pressure were expanded to 15 volumes, the result would be $15 \times 2.75 = 41.25$ cu. ft. One pound of steam thus expanded from 150 pounds pressure will occupy 41.25

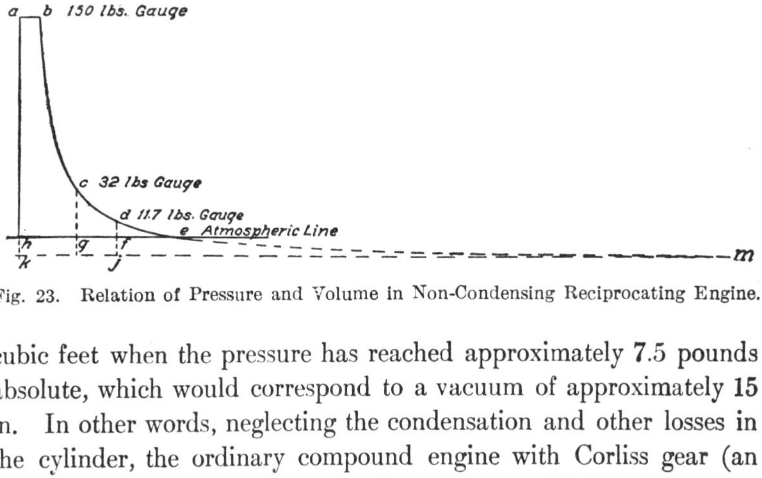

Fig. 23. Relation of Pressure and Volume in Non-Condensing Reciprocating Engine.

cubic feet when the pressure has reached approximately 7.5 pounds absolute, which would correspond to a vacuum of approximately 15 in. In other words, neglecting the condensation and other losses in the cylinder, the ordinary compound engine with Corliss gear (an engine in every way first class), cutting off at one-third stroke, cannot expand steam at 150 pounds boiler pressure lower than to 15 in. vacuum. Any increase of vacuum beyond this point tends only to reduce the back pressure on the piston, and the gain in work is slight, perhaps not enough to pay for the additional work on the air pump, increased size of condenser, and additional circulating water.

Fig. 24 shows, as before, the adiabatic expansion from 150 pounds gauge pressure. If *a b* represents one volume, *h f* would represent fifteen. *h f* will be the back pressure line at 15 in. vacuum, the maximum theoretical work done in the cylinder will be the area *a b d f h*, and the work lost will be the area *h f m k*. Increasing the vacuum below 15 in. gives only a little gain, represented by the area *h f g k*, although more than in the previous case.

A triple-expansion engine will permit of about twenty expansions; that is, the low-pressure cylinder will contain about twenty times the volume displaced by the piston at cut-off in the high-pressure cylinder. In such an engine, the final volume of one pound of steam expanding from the previous pressure will be 55 cubic feet, and the pressure corresponding to this volume would be 5.5 pounds absolute, equal to about 19 in. vacuum. A condenser giving 24 in. vacuum would allow just about difference enough to give a ready flow of steam

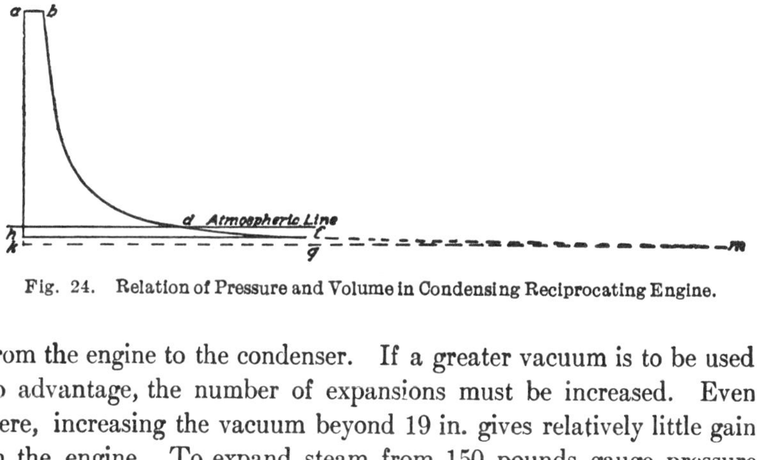

Fig. 24. Relation of Pressure and Volume in Condensing Reciprocating Engine.

from the engine to the condenser. If a greater vacuum is to be used to advantage, the number of expansions must be increased. Even here, increasing the vacuum beyond 19 in. gives relatively little gain in the engine. To expand steam from 150 pounds gauge pressure to 28.5 in. vacuum would require a final volume of 338 cubic feet for each pound of steam admitted to the cylinder, and since one pound at initial pressure occupies 2.75 cubic feet, the steam would have to expand $338 \div 2.75 = 123$ times, approximately. The utter impossibility of such expansion in the triple expansion engine will be evident from the following consideration:

If a triple-expansion engine were to expand the steam to this pressure, with cut-off at one-third stroke, the low-pressure cylinder would have a volume $123 \div 3 = 41$ times that of the high-pressure cylinder, and its diameter would be to the diameter of the high, as 1 is to the square root of 41, or about 6.5. This ratio is not far from three times that found in general practice for such an engine, and about four times that for a compound engine. Assuming that the low-pressure cylinders are now as large as they can conveniently be made, the complete expansion above outlined would require, in the triple-expansion engine, three low-pressure cylinders of the present size. Radiation loss and friction could easily overcome the theoretical gain; to say nothing of the prohibitive cost and weight of the engine.

Consider the diagram in Fig. 25, which shows, as before, the adiabatic expansion between 150 pounds gauge and 28.5 in. vacuum. The black area represents the available work due to the complete expansion of the steam, in excess of that available in the triple-expansion engine, running under 28.5 in. vacuum. This lost energy is

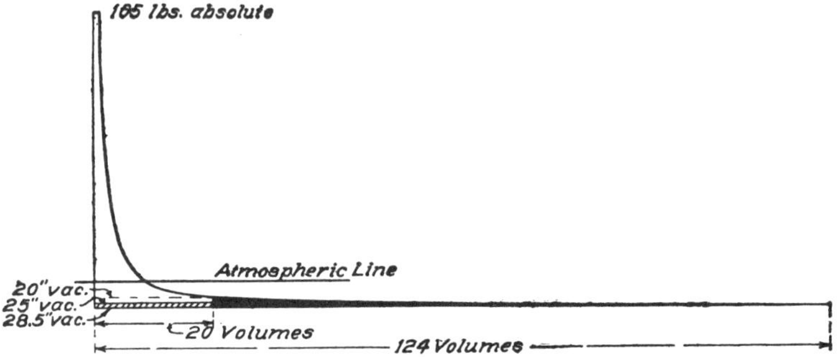

Fig. 25. Energy in Steam not Available for Reciprocating Engine.

about 25% of the total energy available in the steam, or about 35% of the energy available for use in the reciprocating engine with 28.5 in. back pressure. Under the ordinary conditions of 25 in. back pressure, the black area would be augmented by the crosshatched area, making the lost energy about 40% instead of the 35% above. All of this energy is lost by the triple-expansion engine, but can be utilized by the turbine. Low vacuums cause large initial condensation in reciprocating engines, but do not have any disadvantageous effect on the turbine.

The low-pressure turbine can be advantageously used in connection with any reciprocating engine, and their combination will always afford a considerable improvement in economy, and increase the power without increasing the size of the boiler plant. It often happens that engines are operated non-condensing because of the expense of cooling water, and as we have already shown, the relatively small gain would not pay for additional complications and expense, especially if cooling towers have to be provided. The low-pressure turbine, however, will provide enough additional power to pay for the installation of proper equipment. There are already in existence, plants where low-pressure turbines have been installed in connection with engines previously used as non-condensing, and the output has been practically doubled without increased cost for fuel.

It is readily seen from the previous discussion, that even in a plant in which the engines are operated as condensing engines, a considerable gain can be effected by installing a low-pressure turbine, even though using the same condenser facilities as before. In some ways, it is much easier to maintain a high vacuum in such a combination, because the turbine will take the steam at slightly above the atmospheric pressure, and thus prevent a considerable amount of air leakage, which always takes place through the the stuffing-boxes of a low-pressure reciprocating engine.

If saturated steam expands adiabatically from 150 pounds gauge to a pressure of 28.5 in. vacuum, practically half the available energy is developed between the initial pressure and one pound above the atmosphere, and the other half below the latter pressure. It might be said, in explanation, that the work of expansion can be considered as equal to the pressure times the volume; but it is, perhaps, not often realized that the volume of steam will nearly double in expanding from 26 in. vacuum to 28 in., and that, therefore, the available energy is great, although the pressure is low. In most condensing engines, the gain over non-condensing conditions, as determined by actual experiment, does not exceed 30%, even under favorable conditions of steady load. Under average conditions, the gain drops to 25%, and under overload conditions, to a still lower point. In general, a condensing reciprocating engine, if run non-condensing, will carry about 70% of its maximum load, exhausting at, say, two pounds above atmospheric pressure, and, if the steam from such an engine be exhausted into a low-pressure turbine with proper condensing facilities, the latter will develop nearly as much work as was developed by the engine itself, and there will result from the two about 140% of the work which might be expected from the reciprocating engine alone, if run condensing. It is interesting to note that the discussion of Fig. 25 seems to show a possible theoretical gain of about 40% over the engine condensing at 25 in. vacuum, provided the turbine is run at 28.5 in. vacuum.

Fig. 26 shows a study of the possibilities in connection with a Rice-Sargent engine which has been operated for some years in the plant of the General Electric Company at Schenectady, N. Y. This unit operates a 250-v. direct-current generator, and ordinarily runs with a load of 1,200 kw.

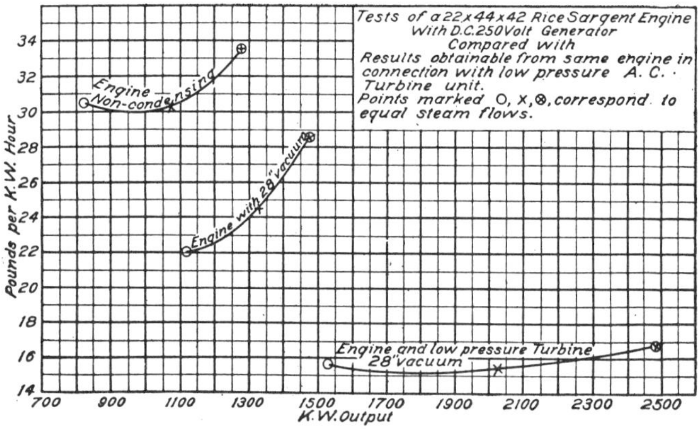

Fig. 26. Curves Showing Economy of Engine with Low-Pressure Turbine.

*The tests were made accurately by weighing condensed steam, the effect of vacuum being determined by holding the steam flow constant, and changing the vacuum. The curves show performance under condensing and non-condensing conditions, and also show what could be accomplished by this engine in combination with a good low-pressure turbine. The rates of gain here shown will seem extraordinary, but they are fairly representative of the possibilities in the average condensing engine plant.

Referring to the curve-sheet, note that the upper curve represents the engine operating non-condensing at 810 kilowatts, the steam consumption being 30.6 pounds per kilowatt. With the load increased to 1,065 kilowatts, the steam consumption is still 30.6 per kilowatt and with the load increased to 1,265 kilowatts, the steam consumption is 33.6 pounds. Operating under these conditions, 1,265 kilowatts is practically the maximum capacity of the unit.

Now, operating condensing with a capacity of 1,140 kilowatts, the steam consumption is 22 pounds per kilowatt; at 1,320 kilowatts, the steam consumption is 24.6 pounds per kilowatt; and operating at 1,470 kilowatts, the steam consumption is 28.8 pounds per kilowatt. Note, however, that the maximum capacity of the unit has been increased from 1,265 to 1,470 kilowatts.

*From a paper by Chas. B. Burleigh, on the "Low-Pressure Steam Turbine."

Now, by the assistance of the low-pressure turbine, vacuum conditions remaining the same, the steam consumption at 1,550 kilowatts is 15.6 pounds per kilowatt; at 2,020 kilowatts, the steam consumption is 15.4 pounds per kilowatt; and at 2,500 kilowatts, the steam consumption is 17 pounds per kilowatt. By this combination, the maximum output of the unit has been increased from 1,265 kilowatts, non-condensing, to 2,500 kilowatts, or from 1,470 kilowatts condensing, to 2,500 kilowatts.

It must not be thought that all this gain can be attained with no compensating loss. In the first place, a surface condenser, to maintain 28.5 in. vacuum, must be about twice the size of one to maintain 26 in., and requires special apparatus that is not only costly, but difficult to maintain. Again, the cost of maintaining a 28.5 in. vacuum is very much more than that of maintaining a 26-in. vacuum, leaving out of consideration the extra cost of condenser and cooling water. After all, it is the dollars and cents that determine the best efficiency, and it is poor economy to obtain the extra power at a greater cost than the returns will warrant. A gain of 35% or more in steam consumption may easily be effected by installing a low-pressure turbine, but the gain in dollars and cents is seldom as great; just what the gain may be, must of course depend upon the local conditions, especially upon the conditions under which the reciprocating engine is operating. In the majority of cases, such installations are worth while, even though used with the usual vacuum.

An interesting application of the low-pressure turbine in connection with rolling mill machinery and other intermittent work, has been worked out by Professor Rateau, and has been made possible by the use of his *steam accumulator*, or *regenerator*. This apparatus regulates the intermittent flow of steam exhausted from the rolling mill engine, let us say, and intended to be used by a low-pressure turbine. The accumulator may consist of a large tank in which are numerous plates over which water can flow, or may contain simply water rapidly circulated by artificial means. As the exhaust steam from the engine enters this accumulator, it spreads out over the exposed water surface, and some of it is condensed if there is an excess of pressure due to more steam being supplied by the exhaust than is being utilized by the turbine. On the other hand, if the turbine utilizes more steam than is supplied by the exhaust, this causes a lowering of the pressure in the accumulator, and a rapid vaporization occurs from the exposed water surfaces, tending to equalize the pressure. The accumulator thus bears the same relation to the transfer of heat from the reciprocating engine to the turbine that a fly-wheel bears to the transfer of work from the cylinder of the engine to mill shafting. Fig. 27 shows one form of the Rateau accumulator. It must be provided, of course, with a safety-valve, set at a pre-

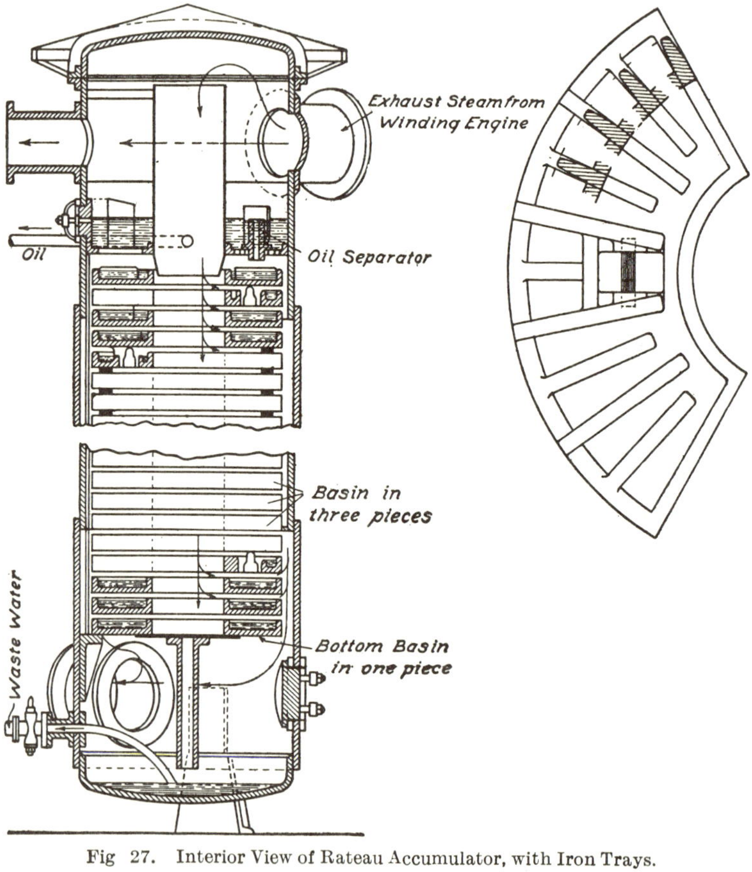

Fig 27. Interior View of Rateau Accumulator, with Iron Trays.

determined pressure, and is usually provided with a reducing valve from the boiler, so that in case the reciprocating engine should stop for a considerable length of time, steam could still be supplied to the turbine through the reducing valve.

The first apparatus of this kind was installed in 1902, and has been very successful. The first to be installed in the United States was at the Wisconsin Steel Company, in South Chicago. In this plant, steam first goes to a receiver to take out the shock due to the puffs of the exhaust. From here it passes to the regenerator. The receiver is fitted with baffle plates and drains for water and oil, by means of which they are thus separated from the steam. This accumulator at South Chicago furnishes steam for a low-pressure Rateau turbine which is used to furnish electric power for general purposes.

Installation. The field of the steam turbine is unfortunately limited in its usefulness by two very important factors; first, its relatively high speed of revolution, even when compounded; and, second, its non-reversibility. If, as in marine work, reversing is absolutely necessary, then another turbine, which runs idle ordinarily, with vanes set in the opposite way must be fitted on the shaft. To make this reversing turbine as small as possible, efficiency is sacrificed, but this is of small consequence, for it is used so little. It of course adds materially to the first cost of the turbine and increases the length of the necessary floor space.

The first and greatest field of turbine usefulness is undoubtedly central station work for the generation of electricity by direct-connected apparatus. It also has an important field in driving blowers, centrifugal pumps, etc., where high speed of revolution is essential. In such cases, it has a distinct advantage, for it may be direct-connected, thus doing away with the belting necessary if reciprocating engines were used. The turbine has been suggested to some extent for driving mill shafting, in which case, of course, the speed is belted down from a small pulley on the turbine to a large one on the countershaft, but this appears to offer no particular disadvantage, for in any case belting would be used, as the countershaft would never be run at the same speed as the ordinary reciprocating engine.

In the field of electric generation the turbine to-day has practically superseded the reciprocator. The number of installations is very great, and probably no new central station is now designed for other than steam turbines. In 1906 the Committee of the National Electric Light Association, after an extensive investigation of turbines, reported a wide use of turbines for electric generation, and their figures showed that about 75% of all the turbine units of 500 kw. or over already installed in the United States, were for electric purposes, and that practically only one new central station abroad had been found installing reciprocating engines. The distinct advantage of turbines for this work is the uniform turning effort, the high speed of rotation permitting the use of a very much smaller generator, and the smaller floor space, requiring less capital outlay in land and engine-house. These features place it in striking contrast with the ponderous slow-moving Corliss engine.

The General Manager of the Metropolitan Street Railway Co. of Kansas City is authority for the statcment that in that station six 5,000 kw. units of a well-known make of turbine could be installed in space previously occupied by three 3,000 kw. engine-driven

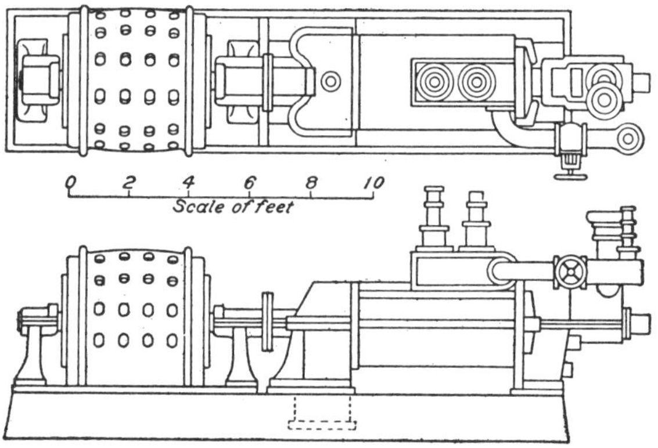

Fig. 28. Plan and Elevation of 500-K. W. Westinghouse Turbo-Generator. This is Same Scale as Fig 29; Notice Difference in Space Required.

units. Or in other words, 30,000 kw. of turbine power could be put into a building where before only 9,000 kw. of engine power had been possible. This probably is greater than would ordinarily be met with, but the difference in any case is large, the saving in space depending upon the type of turbine. The average horizontal turbine and generator with auxiliary apparatus will occupy about three-fifths of the space needed for a slow-speed, engine-driven generator of the same power, and a vertical turbo-generator somewhat less space than the horizontal.

A further distinct advantage of the turbine is in the fact that, since there are no valves to adjust, the efficiency can be lowered only by wear, and then only slightly; on the other hand, in reciprocating engines, if the valves are not set exactly right, very poor economy will result, and the opportunities for wear are far greater than in turbine engines. Again, the turbine can use high degrees of superheat because there is no lubricant to burn; there is also little danger of entrained moisture in the steam wrecking the turbine, and the

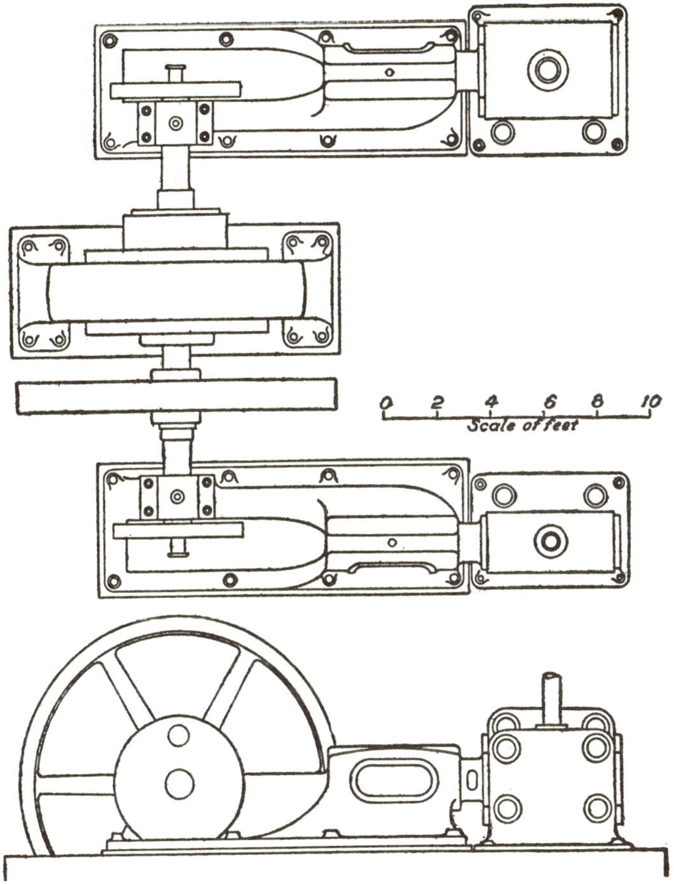

Fig. 29. Plan and Elevation of 500-K. W. Corliss Engine-Driven Generator Set. Compare with Fig. 28.

absence of oil in the condensed steam greatly lessens trouble in the boiler if the condensation is used for feed water. The economy of space was graphically illustrated by Fig. 1, and Figs. 28 and 29 tell the same story but with different types of engine and turbine.

Figures showing the relative space occupied by reciprocators and turbines are of little value unless the size of condenser and condensing auxiliaries are taken into consideration, for, as before mentioned, they may easily be, in the case of the turbine, twice the size of those used with a reciprocating engine of the same power. The apparent saving of space, therefore, may be offset by these auxiliaries. By placing the condensers underneath the turbine, as is frequently done at the present time, not only may a considerable amount of floor space be saved, but the turbine can more readily exhaust into the condenser. As we have already seen, at high vacuum the volume of steam is very large, and the exhaust pipe from the turbine will be proportionally large. It would thus appear that to have the condenser any great distance from the low-pressure end of the turbine would be not only a distinct disadvantage, but offer a considerable practical difficulty.

Turbines, as we have seen, require very much smaller foundations than reciprocating engines of the same power, and these foundations will therefore cost very much less. It is hard to get a direct comparison between turbines and reciprocating engines as a class, because the foundations for high-speed reciprocating engines will not be as massive as for the heavier, low-speed engines. The turbine, occupying less floor space, will require smaller buildings and less land, and this will in a number of cases be a substantial saving in first cost and subsequent interest charges.

So far as the first cost of a generating plant goes, there is at the present time very little difference between those using reciprocating and those using turbine engines. The turbine itself costs more than the reciprocating engine of the same power, but on the other hand the generator for the turbine costs very much less. Again, the condenser and pumps, if high vacuum is to be maintained, will cost two or three times as much as for the reciprocating set, while the cost of erection is decidedly in favor of the turbine. It is not easy to get a direct line on the relative cost of turbine and engine installations, for the figures available appear to vary about as much between reciprocating engines and turbines as might be expected to be found between various installations of reciprocating engines, and undoubtedly turbine installations in some cases cost relatively more than in others. It seems probable that the cost of the turbine is regulated more by the cost of the reciprocating engine with which it has to compete, than by the actual cost of manufacturing the turbine. All in all, there is likely to be a somewhat less cost of complete installation in favor of the turbine, but the difference will not be large in any case, and in powers under about 100 kw., it is probable that the engine installation is fully as cheap. This does not take into account the value of land and buildings, which in all cases is an important factor in favor of the turbine.

Performance. The losses occurring in the steam turbine consist principally of loss of velocity of the steam itself due to friction in contact with the vanes and guides; friction of the disks revolving through a chamber filled with steam; eddying of the steam jet, due to improper speed of the revolving disks; radiation; bearing friction. The two latter items are not large, and under ordinary conditions would consume less than 2% of the power.

The most important losses are due, first, to the friction of the steam jet against the vanes and guides, which will be approximately proportional to the cube of the velocity of the steam relative to the vanes or guides, and second, to the considerable amount of friction of the disks as they revolve in the chamber filled with steam. This friction generates heat which raises the temperature of the steam and metal parts and thus causes the re-evaporation of some of the condensed moisture. Since this adds some heat to the expanding steam, the expansion is not absolutely adiabatic. The smoother the revolving wheels are made, the less will be this friction, a fact well illustrated by a reported improvement of about 1% in steam consumption which was effected in a well-known make of turbine by making the riveting on the revolving disks perfectly flush. To these losses may properly be added the generator losses which, of course, are a factor of the speed of revolution.

With either reciprocating engines or turbines, the steam economy is much better in large than in small units, and especially is this true of the turbines of the reaction type. In small turbines of this type, the steam friction is high and the leakage large, and this makes it undesirable to build this type of turbine in sizes much below 500 kw. For the impulse type of turbine, these losses are not as important in the smaller powers, and DeLaval, Curtis, and Rateau turbines of comparatively small power can be built to give nearly as good steam economy as larger turbines of the same type, and can easily excel small reciprocating engines.

The steam consumption of the turbine depends naturally enough upon the vacuum, steam pressure, degree of superheat, variation in load, and variation in speed. It has already been explained that the turbine can utilize the lower ranges of vacuum far better than can the reciprocating engine, but it could not, in all probability, use the higher pressure ranges with as good economy as the best reciprocating engines. If the turbine runs at a vacuum of 27 in., its steam consumption will be practically on a par with that of the reciprocating engine, and it will show a gain of about one-half pound of steam per kw.-hr. for each extra inch of vacuum, below 25 in. But from the saving effected by this one-half pound of steam must be deducted the extra cost of maintaining the high vacuum, if the real economy is desired. Not only can the turbine theoretically utilize the greater vacuum to better advantage, but it has an advantage also in a practical way, because with the reciprocating engine, a very high vacuum cools the cylinder walls and thus causes a relatively large initial condensation, which difficulty is not met with in the turbine, the high vacuum having no detrimental effect. It thus has both a theoretical and practical advantage.

Superheated steam, whether used in the reciprocating engine or in the turbine, will reduce the steam consumption; but in the reciprocating engine, superheating cannot be carried very high, as the cylinder lubricant is likely to be burned, and there will be little condensation in the cylinder to help out the lubrication. The turbine is not handicapped in this way, but nevertheless high degrees of superheat are likely to cause trouble due to unequal expansion in the casing, the temperature at the high-pressure end being so much greater than that at the low-pressure end. This expansion is troublesome, but should be provided for in the design.

Superheat affects the economy of the steam engine in two ways; it carries additional heat units into the cylinder, and lessens condensation. It also helps in the turbine in two ways; it carries additional heat into the turbine, and, being less dense than saturated or moist steam, causes less friction within the turbine, and thus effects a mechanical as well as a theoretical gain. It is generally reported that the gain is 10% for each 100° of superheat, but tests which appear to be thoroughly reliable do not seem to bear out this claim. $7\frac{1}{2}\%$ to 8% is a better figure.

The saving in steam will be from 1.5 to 1.75 pounds per kw.-hr. for each 100° of superheat, but the real economy resulting from this superheat will be the difference beween the value of this saving in steam and the cost of superheating. The superheating plant costs more, not only for the additional expense of the superheater, but for piping, valves, etc. Cast-steel fittings, and valves with nickel-steel valve stems, are usually required for high degrees of superheat.

The usual steam pressure in turbine work is about 150 pounds gauge. If lower than this, some gain in steam consumption may be had by an increase in boiler pressure, but an increase over 150 pounds does not appear to be productive of great economy. A reference to Fig. 25 will readily show that increasing the pressure above 150 pounds will add very little to the area of available work. Fig. 30 shows the curves of economy of a 30-H. P. turbine at different steam pressures. The gain is less and less the higher the pressure becomes, and is small from 75 to 100 pounds. From 35 to 100 pounds the gain is about $33\frac{1}{3}\%$, but this gain is not due entirely to the rise in steam pressure.

The study of steam nozzles has shown that to use steam efficiently, the nozzle must be properly designed with reference to both the initial and final pressures. Now, if the nozzle on this turbine were designed for 100 pounds pressure, it could neither utilize steam economically at 35 pounds, nor at 150 pounds pressure. To show the real gain due to an increase in steam pressure, it would be necessary to have nozzles in each case that were designed for the specific pressures used. Then, and only then, would the curves show the true gain due to increase in pressure. But a study of Fig. 25 shows that if the theoretical gain is small the practical gain cannot be large. It must, moreover, be borne in mind that a high-pressure plant costs more than a low-pressure plant, and for stationary work very high pressures will not pay. On shipboard, where space and weight are at a premium, it may be good engineering policy to install very high pressures, even though the first cost is greater.

Fig. 31 shows the curves for a 600-kw. Curtis turbine with varying pressures. In this type of turbine, the same conditions exist as in the previous one, the nozzles being designed for only one pressure.

The economy of a turbine varies with the load, as does the economy of the reciprocating engine, but not perhaps to as marked an extent, and the economy depends of course upon the type of turbine. Turbines like the DeLaval and Curtis admit steam through a number of nozzles which are opened and closed either automatically by the governor or by hand. At normal load, about two-thirds of these nozzles would be open and a 50% overload could then be carried with all nozzles open. In the Parsons turbine, steam is admitted all around the circumference of the drum but the admission is intermittent. For heavy loads the valve remains open for longer in-

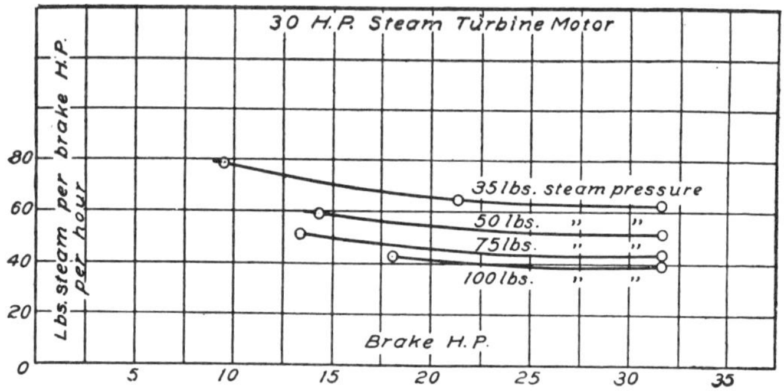

Fig. 30. Curves Showing Economy of 30-H. P. Turbine.

tervals, and when the load is such that the valve remains open all the time, further overloads can be provided for only by resorting to a by-pass which admits high-pressure steam to the second stage of the turbine. In such cases, of course the economy falls off, for the steam does not get the benefit of full expansion. At low loads, there is not a great deal of choice between the different types of turbine, but those that can carry a large overload without opening a by-pass are bound to be the most economical under these conditions.

Overload is taken care of in a reciprocating engine by increasing the cut-off, but, as this reduces the number of expansions, this method is uneconomical. For small ranges of load, the relative economy of turbine and reciprocator are not very different, but the effective range of the turbine is much greater than for the reciprocating engine.

A good turbine will carry 100% overload for a short time and will carry 50% to 60% overload on approximately 10% more steam. Fig. 32 shows characteristic curves of steam consumption at varying loads.

A variation in speed of the turbine within moderate limits does not materially affect the economy. The best speed of the vanes (see Page 17) is half the velocity of whirl ($\frac{1}{2}$ V cos α). Moderate departures from this speed do not materially affect the economy, provided the entrance angles of the vanes are such that the steam jet strikes without shock. The angle of the vane must depend upon the speed, and once fixed, any variation in speed must of course cause the steam jet to spatter and form eddies, a source of material loss. This is entirely apart from the question of whether or not the designed speed of rotation is the most economical. To avoid spattering and eddy losses, the vane angle must change with the speed, which is evidently impossible

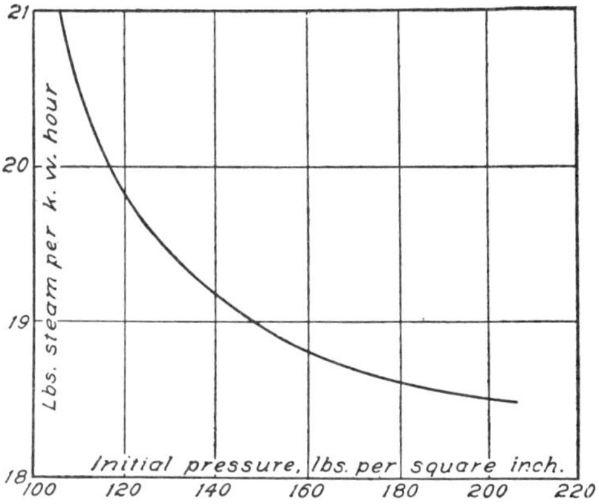

Fig. 31. Economy Curves of 600-K. W. Curtis Turbine with Different Steam Pressures.

A rapid change in load will cause cylinder condensation in a reciprocating engine, so that, on test under steady load, the engine is likely to show up better than it would under service conditions. With the turbine, this is not so. Here there is no such condensation, and the performance under test is far more likely to agree with performance under service conditions. Both types of motor will fall off under service conditions, but if an engine and turbine do equally well under test, under such widely varying conditions as exist in a central station, for instance, the turbine ought to show up better in actual service.

A reciprocating engine is usually designed for a low average load and, therefore, it will permit a relatively large increase in load, but it is generally working on a slight underload, and hence at less than the maximum efficiency. The turbine, on the other hand, is usually designed for its normal and most economical load, taking care of overload by opening more nozzles at theoretically the same efficiency, or by opening a by-pass at somewhat less efficiency. This should give the turbine a still further advantage at the end of the day's work.

Tests. Tests of reciprocating engines usually give steam in pounds per indicated horse-power per hour, but there being no

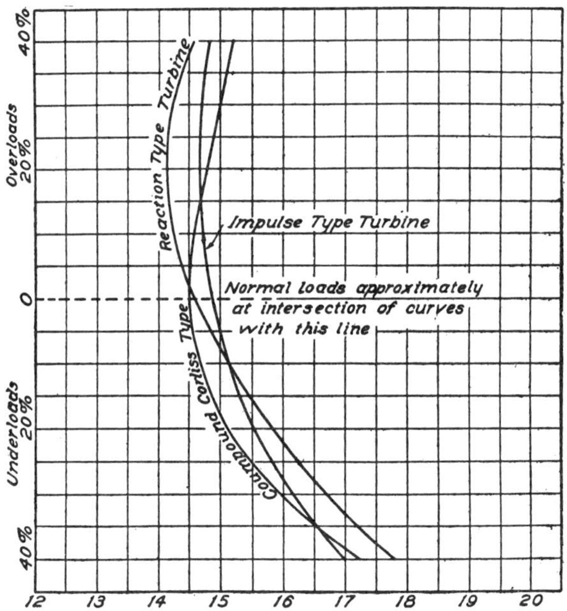

Fig. 32. Economy Curves of Turbines and Compound Corliss Engine with Varying Loads.

indicated horse-power for a turbine, the comparison must be made on some other basis. Brake, or shaft horse-power may readily be obtained for a turbine, and in engine tests, where the brake horse-power has been determined, there is of course opportunity for a direct comparison. However, since engineers are in general more familiar with steam rates per I. H. P., it seems well to consider how a *comparative* I. H. P. may be had for the turbine. Various tests to determine the relation between brake and indicated power on reciprocating engines seem to show that 92% is a fair figure for a good engine. 92% then of the steam rate per brake horse-power would give the rate per *comparative indicated* horse-power.

The largest field for the steam turbine being central station work, it follows that by far the larger number of turbine tests are quoted in terms of electrical units. It is costly to fit a brake for a large turbine and entirely useless when the power delivered at the switchboard can be read off at once. For electrical work, of course reciprocating engine tests are often quoted in the same electrical units, in which case, there are abundant opportunities for direct comparison.

Suppose, however, that it is desired to compare steam per I. H. P. with a corresponding rate per kilowatt-hour at the switch-board. 1 kw. = 1.34 electric horse-power measured on the switch-board, which is evidently shaft or brake output less losses in the generator. Since the efficiency of a good generator is not far from 95%, the brake horse-power will be equal to the electric horse-power divided by $\frac{95}{100}$. We may say, therefore, in ordinary cases, that

$$\text{B. H. P.} = \frac{1.34 \times \text{kw.}}{.95}$$

since we assume, that $\frac{\text{B. H. P.}}{\text{I. H. P.}} = \frac{92}{100}$,

we have $\text{I. H. P.} = \frac{1.34 \times \text{kw.}}{0.95. \times 0.92} = 1.53$ kw. approximately.

Steam per kw.-hr. then, divided by 1.53 would give the steam per *comparative indicated* horse-power per hour, or

steam per I. H. P.-hr. $\times$ 1.53 = steam per kw.-hr.

A turbine using 20 lbs. of steam per kw.-hr. would be about on a par with a reciprocating engine using $\frac{20}{1.53}$ = 13 lbs. per I. H. P.

In comparing the performance of one engine with the performance of another, or one turbine with another, or an engine with a turbine, pounds of steam per horse-power per hour is generally the rough basis of comparison, but this is very crude and often misleading. For instance, one test may be made with superheated steam and another with saturated or even moist steam, or one may have a higher steam pressure, or the vacuums may be different.

To get an approximately intelligent comparison, all tests should be reduced to a standard degree of superheat, pressure, and vacuum, or better still, if the comparison is between two, correct both to the average conditions of the two. The corrections applied are more or less arbitrary, and it is manifestly unfair to apply them all to either test. If each is corrected for half the difference, a much more reliable comparison is likely to result.

It is generally accepted that the steam consumption will decrease about 8% for each 100° of superheat, about 5% for each inch of vacuum below 28 in., and about 5% for 50 lbs. rise in steam pressure between 100 and 150 lbs., and 3% for similar rise between 150 and 200 lbs. The manufacturer usually gives guarantees of steam rates for various pressures, vacuums, and degrees of superheat. When such figures are available, it is probable that their use would lead to more satisfactory results than if the rough approximations mentioned above were used, but such figures would be correct only for the one individual turbine, and in the large majority of cases the engineer is compelled to use the approximations. They are in most cases fair and satisfactory in the absence of definite data.

To illustrate this method, consider a turbine at 177.5 lbs. (gauge) steam pressure, vacuum 27.3 in., superheat 96° F., consuming 15.15 lbs. steam per kw.-hr., and another using 179 lbs. steam pressure, 29.5 in. vacuum, and 116° F. superheat, consuming 13 lbs. of steam per kw.-hr. The average conditions are 178.2 lbs. steam pressure, 28.40 in. vacuum, and 106° F. of superheat.

The work will appear clearer if arranged in tabular form as in Table I.

TABLE I

Steam Consumption Tests

	Turbine #1	Turbine #2	Average Conditions	Turbine #1 Correction %	Turbine #2 Correction %
Steam by Gauge	177.5 lbs.	179 lbs.	178.2 lbs.	0	0
Vacuum	27.3 in.	29.5 in.	28.4 in.	−5.5%	+5.5%
Super-heat	96°F.	116°F.	106°F.	−0.8%	+0.8%
Steam per Kw.-Hr. Observed	15.15 lbs.	13 lbs.		−6.3% or	+6.3% or
Steam per Kw.-Hr. Correct'd	14.19 lbs.	13.82 lbs.		−0.96 lb.	+0.82 lb.

The correction for steam pressure, being only for .7 lbs., is too small to be of consequence in this case. The vacuum correction is ± 1.1 inches, and at 5 % per inch (the decrease in steam consumption for each inch of vacuum, as explained on Page 52), the correction would be $\pm 5\frac{1}{2}\%$. The superheat correction is for 10°, or, as the decrease in steam consumption for 100° of superheat is 8%, this will be $\frac{1}{10}$ of 8% = 0.8%. The sum of these corrections gives $\pm 6.3\%$, making .96 lbs. to be subtracted from turbine #1, and .82 lbs. to be added to turbine #2. The final steam consumptions, then, which should be compared are 14.19 lbs. and 13.82 lbs. instead of 15.15 lbs. and 13 lbs. Turbine #2 appears, therefore, to use about 3% less steam than turbine #1 under similar conditions.

Another and perhaps more satisfactory method of comparison is by means of the heat units used. This computation may be made readily from the steam tables. Using the same tests as given above, turbine #1 uses steam at 177.5 lbs. gauge pressure = 192.2 absolute, at which pressure each pound of dry saturated steam contains 1197 B. T. U. If we allow $\frac{1}{2}$ B. T. U. for each degree of superheat, then, for 96° F. we should add $96 \times .5 = 48$ B. T. U., and each pound would then contain 1245 B. T. U. at admission. If this steam is condensed at a pressure of 27.3 in. vacuum = 1.33 lbs. absolute, each pound of the condensation will contain 80 B. T. U. which will be returned to the boiler in the feed water. The net amount then consumed by the turbine and carried away by the cooling water of the condenser is $1245 - 80 = 1165$ B. T. U. per pound. 15.15 lbs. would represent $15.15 \times 1165 = 17,650$ B. T. U. per hr. or 294 B. T. U. per kilowatt per minute.

Turbine #2 uses 13 lbs. of steam at 179 lbs. gauge pressure and 116° F. superheat, condensing at 29.5 in. vacuum. In this case, each pound of dry steam at admission would contain 1197.3 B. T. U. and 116° F. superheat would add about 58 B. T. U. more, making 1255.3 B. T. U. per pound. Condensing at 29.5 in vac. = .25 lbs. absolute, each pound of condensed water would contain 27 B. T. U. to return to the boiler in feed water, leaving $1255.3 - 27 = 1228.3$ B. T. U. to be used by the turbine. 13 lbs. would represent $\frac{13 \times 1228.3}{60} = 266$ B. T. U. per min. to compare with 294 in the previous case.

Here, again, the direct comparison is likely to be misleading, unless due account is taken of the difference in conditions. The gain is apparently about 10% in favor of turbine #2 on the heat unit basis taken under the actual working conditions of each, but the fact must not be lost sight of, that turbine #2 is working under more favorable conditions of vacuum and ought to show a much better efficiency. It appears from this discussion that both turbines work under the conditions of design with but little difference in actual economy.

Turbine manufacturers are in the habit of reporting tests of the turbine only, no account being made of the auxiliary apparatus. This is manifestly misleading, for with a 29-in. vacuum, the power consumed by auxiliaries may easily be twice what it would be for a 27-in. vacuum. This extra power and the cost of maintaining it in a measure goes to offset the gain due to the higher vacuum.

"ECONOMY" STEAM TURBINE OPEN FOR INSPECTION

Courtesy of Kerr Turbine Company, Wellsville, New York

STEAM TURBINES

PART II

COMMERCIAL TURBINES

*In this description of commercial turbines it will be convenient to classify them as follows:

I. IMPULSE TURBINES
- Single-Stage Type
- Compound Type
  - Compounding by Velocity Steps only
  - Compounding by Pressure Stages only
  - Compounding by both Velocity Steps and Pressure Stages

II. REACTION TURBINE

III. COMBINED IMPULSE AND REACTION TURBINE

IMPULSE TURBINES

SINGLE-STAGE IMPULSE TURBINES

Probably the simplest type of turbine is the one with a single stage, that is, a single set of nozzles and a single rotating wheel, but, as already pointed out, the velocity of rotation in a turbine of this sort is usually so great that some device must be employed to reduce the rotational velocity. This may be done in two ways.

As has been previously stated, the feature of importance is not the rotative speed but the peripheral velocity of the wheel, which is somewhat less than one-half the steam velocity. Maintaining this peripheral velocity constant, turbine rotors of comparatively small diameter may be used, the high rotative velocity being reduced *by means of gearing;* or, *the diameter of the turbine rotor may be increased,* the rotative speed thereby being reduced in the same ratio that the diameter of the wheel is increased.

---

*Many writers group by themselves all turbines using buckets of the Pelton type, but this does not seem to be a proper classification, as it is the action of steam in the turbine that makes it belong to a certain type, and not the style of bucket that is used. Turbines using Pelton buckets may belong to any of the impulse groups.

Both of these methods have been employed in turbines which have been put on the market, the first method being characteristic of the De Laval turbine, and the second, of the earlier forms of the Riedler-Stumpf machine. The manufacturers of the latter discarded this scheme in their later designs in favor of a compound turbine of some sort.

De Laval Turbines. The turbine designed and developed by Dr. Gustav De Laval of Sweden was among the first to be commercially successful. His first turbine, which was used to run the famous De Laval cream separators, was of the pure reaction type, similar

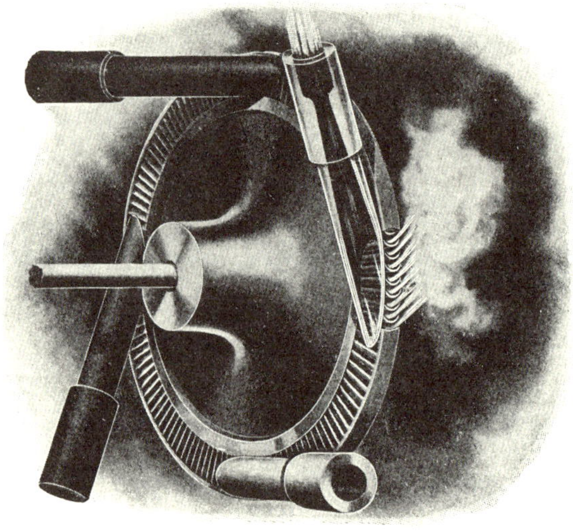

Fig. 33. Principle of Operation of De Laval Steam Turbine

in action to the old Hero engine. This turbine was not economical in steam consumption, but, as it was used for very small powers only, this factor was not important, and commercially, the machine was very successful. This success led to the desire to build larger turbines, and in developing them the reaction principle was abandoned.

The essentials of the motor element of the De Laval turbine are illustrated by their familiar trade-mark, shown in Fig. 33. They consist of a rotating disk, having vanes on its periphery; a number of nozzles in which the steam is expanded from boiler pressure to the pressure in the exhaust chamber and delivered in a jet against the vanes; a shaft to which the rotating wheel is fixed, so arranged that at high speed the rotating element can revolve about its own center of gravity* instead of its geometrical center; and a set of reducing gears to reduce the high rotative speeds to the desired amount. It is an impulse turbine with a single wheel carrying one row of buckets, and is a single-stage turbine in all respects. The steam is directed against the vanes from nozzles with flaring sides, which are so designed as to give it the maximum velocity and to expand it within the confines of the nozzle to the pressure in the exhaust chamber, thus transforming all of the heat energy of the steam into kinetic energy. The nozzles deliver the steam jets at the side of the wheel, and for a maximum efficiency should make as small an angle as possible with the plane of rotation.† In the De Laval machine this angle is 20 degrees.

For small turbines, the entrance and exit angles of the vanes are 32 degrees, increasing to 36 degrees for the larger sizes. Under these conditions the best peripheral velocity will be about 1900 feet per second while the velocity of the steam issuing from the nozzles is 4000 feet per second. In most impulse turbines the peripheral velocity varies from 1400 in the larger sizes to 500 in the smaller sizes. These speeds are high, even for turbine work, and necessitate the solution of very interesting engineering problems. These velocities, with the diameters used for De Laval machines, mean revolutions of about 10,600 per minute in the larger sizes and 30,000 per minute in the smaller sizes, these speeds being reduced by helical gears to approximately 900 and 3000 revolutions per minute, respectively.

In the small-sized turbines this gearing consists of a pinion and a single gear, but in the larger-sized turbines there is a single pinion with a gear on each side. This method has the advantage of distributing half the load on each gear, thus lowering the pressure on the teeth and eliminating the side pressure on the bearings of the flexible shaft.

Nozzles. The nozzles are set in the casing in which the wheel is enclosed and are opened and closed by means of hand valves. A

---

*Lack of uniformity in the density of the steel might cause the center of gravity of the wheel to be outside of its geometrical axis.

†Page 17, Part I.

detail of the nozzle and valve is shown in Fig. 34. *A* is an annular space in the casing acting as a steam chest, *C* is the valve which permits opening or closing of the nozzle, and *B* is the nozzle itself. The nozzle is fitted into a taper hole in the casing and drawn into place by a nut.

The design of the nozzle naturally depends upon the pressure used, the degree of superheat, and the vacuum or back pressure. The nozzles being easily removed, it is apparent that a turbine can readily be altered to meet different conditions by inserting new nozzles. A condensing turbine is often equipped with an extra set of nozzles designed for non-condensing conditions, which may be used with better economy in case the vacuum fails.

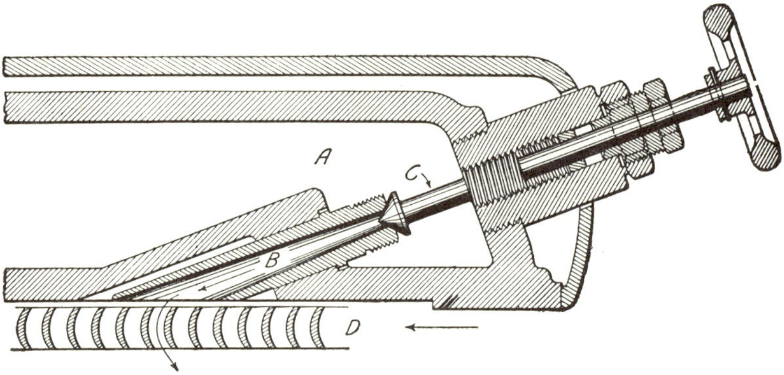

Fig. 34. Detail of Nozzle of De Laval Turbine

There are usually from 2 to 24 nozzles in the casing, and the power developed at any time naturally is proportional to the number of nozzles in operation. The clearance between the wheel and the nozzle is about $\frac{1}{8}$ of an inch. The clearance between the tips of the blades and the casing is not a matter of importance, for there is no tendency to steam leakage, the pressure in all parts of the casing being practically the same as the back pressure. This clearance, therefore, may be whatever practical conditions require. Fig. 35 is the exterior view of the turbine and generator, showing nozzles and valves set in the casing. By inserting nozzles in the holes which are shown plugged in the figure, a greater power can be obtained.

Vanes. The vanes are of the crescent shape common in impulse turbines. They are made of drop-forged steel which resists erosion

Fig. 35. De Laval Double-Geared Turbine Driving Direct-Current Generator of Twin Magnet Frame Type
Courtesy of De Laval Steam Turbine Company, Trenton, New Jersey

and have bulb shanks, as shown in Fig. 36, which are driven into place. The outer ends of the vanes fit close together, thus forming a continuous ring, which prevents any movement at the ends of the vanes.

Steam at high velocities, especially if wet, is liable to cause appreciable wear on the vanes, the wear being practically all on the entrance side; but it is not very great, and tests of a 100-horsepower turbine have shown that wear on the buckets could be as great as $\frac{1}{16}$ of an inch without increasing the steam consumption more than 3 per cent, according to the report of the manufacturers.

Wheel. At the very high speed employed, centrifugal forces are enormous, hence, special high-grade nickel steel must be used in the manufacture of the rotating elements. The steel is said to be high in carbon and to possess a tensile strength of approximately 135,000 pounds per square inch. The wheel is shown in cross section in Fig. 37 and is designed to be of uniform strength throughout, except just below the rim, where a narrow annular groove is turned purposely to make this section weak, for the following reason:

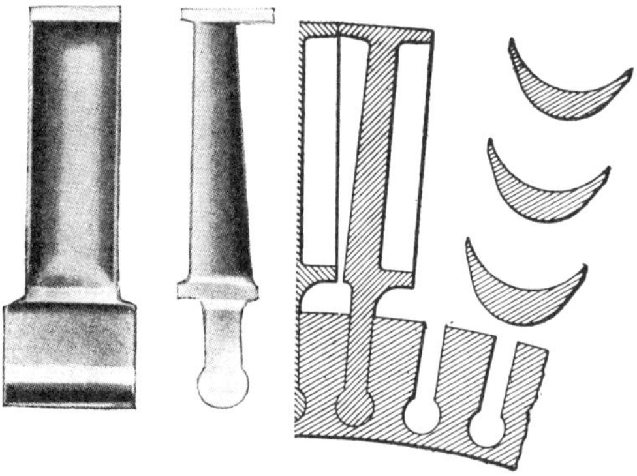

Fig. 36. Drop-Forged Vanes and Method of Attachment in De Laval Turbine

Centrifugal force increases as the square of the speed, and should the safety devices fail to work, the rotating wheel must ultimately burst. The reduced section near the periphery of this wheel makes the stresses at this point approximately 50 per cent greater than elsewhere, and yet, at normal speeds, this will be perfectly safe, as the factor of safety is between 4 and 5. Now, since the centrifugal force increases directly as the square of the number of revolutions, the stresses at the weakened point, when the speed is double, will be four times as great, that is, about equal to the ultimate strength of the material. Consequently the rim will burst and fly into many small pieces, doing but little damage, however, as the casing is made heavy enough to restrain these fragments. When the rim flies off, the stresses in the main portion of the wheel are thereby greatly reduced and no further damage can ensue. Wheels without this weak section have burst under experimental tests into a few large pieces possessing enough energy to break through a 2-inch cast-iron casing.

On each side of the wheel are hubs extending into cylindrical openings in the casing. These are known as *safety bearings* and work with slight clearance under ordinary conditions. Should the rim burst, the wheel would at once become unbalanced and the result-

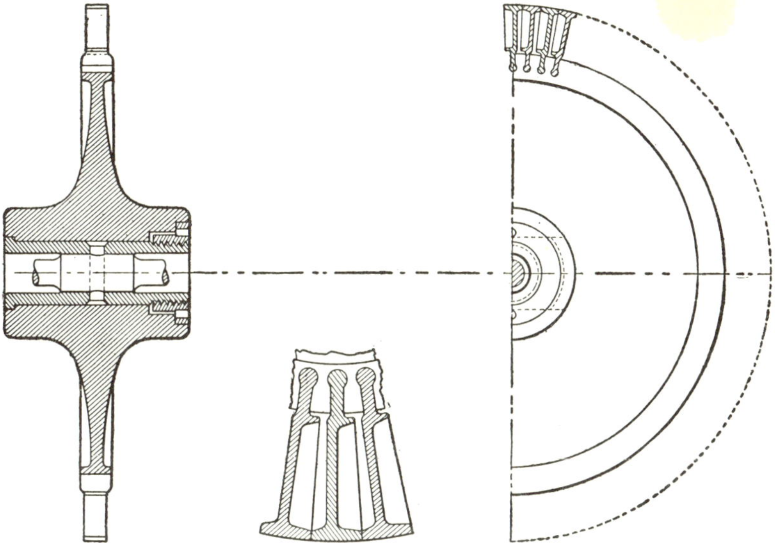

Fig. 37. Method of Mounting Wheel of Small De Laval Turbine

ing eccentricity of the center of gravity would cause the wheel and shaft to rotate off-center, bringing a considerable pressure of the hub against these safety bearings. These acting as a brake, together with the absence of further impelling forces due to the loss of rim and buckets, will quickly bring the rotating wheel to a stop.

For small wheels a bushing is fitted and shrunk to a short swelling on the shaft and, in addition, is pinned in place. The hub of the wheel is bored to fit this bushing and, together with the shaft, is drawn into place by a nut, as shown in Fig. 37. The wheel may readily be removed from the shaft by loosening the nut.

For large wheels such a construction is not desirable, because a wheel with a hole in the center is not nearly so strong as one without such a hole, and in the larger sizes of turbine the strength of the wheel is an exceedingly important factor. The hub, therefore, in such a wheel is solid, but is recessed to fit the flanged end of the shaft, as shown in Fig. 38. The recess is tapered $\frac{1}{2}$ inch to the foot, to fit the shaft, which is securely bolted in place, as shown. The rim of the wheel is drilled parallel to the shaft, with cylindrical holes milled out,

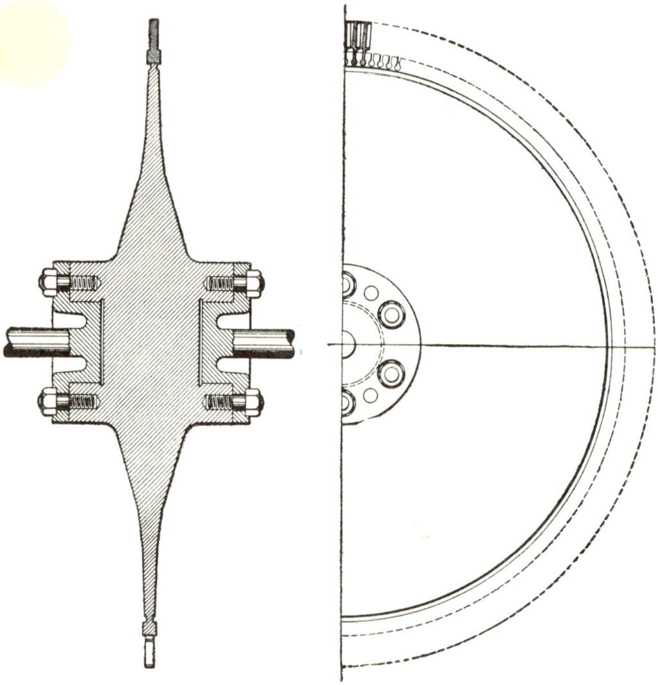

Fig. 38. Method of Mounting Wheel of Large De Laval Turbine

as shown in Fig. 36, to hold the bulb shanks. This makes a strong construction, and the vanes are easily replaced when necessary.

Shaft. When a body is rotating at high speed, it must be very carefully balanced by distributing the material symmetrically about the center of rotation. If the center of gravity of the rotating mass is not absolutely at the center of the shaft, a vibration more or less serious will be set up, because a rotating body tends to rotate about its own center of gravity instead of its geometrical center, thus causing a pressure alternately on one side or the other of the bearing. For speeds of 3000 r.p.m., which are common in compound turbines, the wheels can be balanced on knife-edges, the wheel disks being drilled at certain points until they become perfectly balanced. It is claimed that careful work in this matter will ensure the center of

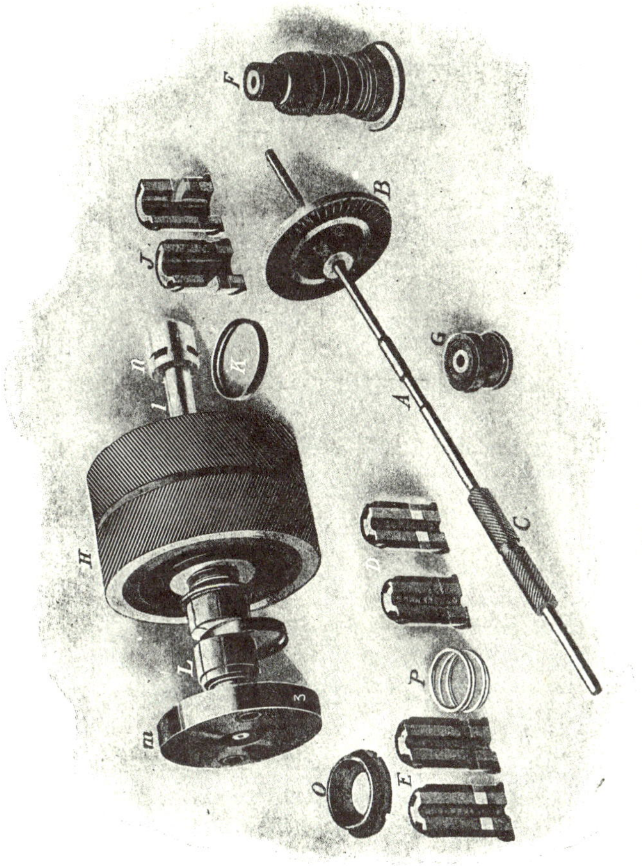

Fig. 39. Working Parts of De Laval Turbine

gravity of the wheel being within $\frac{4}{1000}$ of an inch of the geometrical center. Small as this error may be, it would be prohibitive at the high rotative speeds used in the De Laval turbine; hence the adoption of the long, slender shaft on which the wheel is mounted. This bends slightly, and allows the wheel to rotate about its own center of gravity without vibration. This feature is distinctive of the De Laval machine. The relatively small diameter of shaft is astonishing, being only a little over $1\frac{1}{4}$ inches at its smallest section for the 300-horsepower turbine.

Gears. The speed-reducing gears are in the ratio of about 10 to 1; i.e., if the turbine rotor has a speed of 30,000 r.p.m., the larger gears have a speed of 3000 r.p.m. At the desired place, a swelling on the shaft is provided in which the pinion teeth are cut. In the smaller sizes only one large gear is used, but in the larger

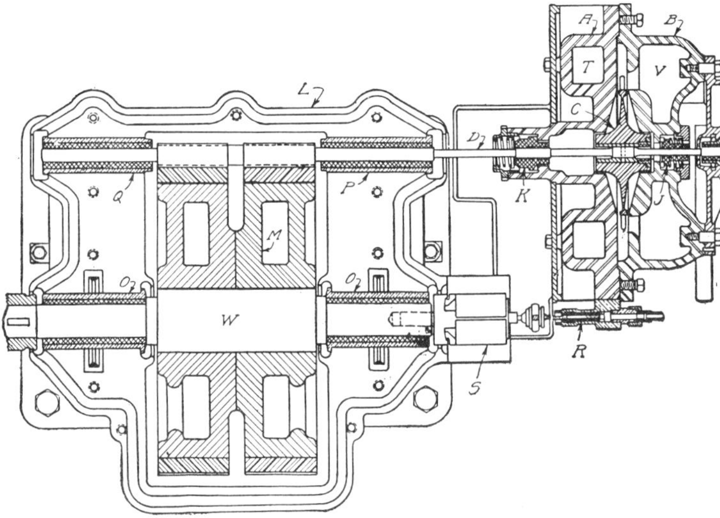

Fig. 40. Horizontal Section of De Laval Single-Geared Turbine

A, Wheel Case; *B*, Wheel Case Cover; *C*, Turbine Wheel; *D*, High-Speed or Pinion Shaft; *F*, Outboard Bearing Bracket; *I*, Outboard Ball-Seated Bearing; *J*, Outer Packing Bushing; *K*, Inner Packing Bushing; *L*, Gear Case; *M*, Gear; *O*, Gear Shaft Bearings; *P*, Inner Pinion Bearing; *Q*, Outer Pinion Bearing; *R*, Vacuum Governor Air Valve; *S*, Governor; *T*, Nozzle Chamber; *V*, Exhaust Chamber; *W*, Gear Shaft.

machines there are two large gears, one on each side of the pinion. The teeth are cut spirally at an angle of about 45 degrees, as shown in Fig. 39, and have double sets of teeth at 90 degrees to each other.

These reduction gears are fine examples of engineering and mechanical skill, as only the best work would stand up under the high speeds of rotation. The shaft on which the pinion is cut is of nickel steel, but the gears are made of soft steel, low in carbon. They have a peripheral velocity of about 100 feet a second and, if kept free from grit, will run for a long time with little or no wear. These gears were originally made of bronze, but this metal proved unsatisfactory because of the crystallization which it developed, and which resulted in the fracture of the teeth after a few years' continuous use.

Fig. 39 illustrates the various working parts of the De Laval steam turbine. *B* is the rotating bladed wheel, *A* the long flexible shaft, *C* the pinion cut on the shaft, *H* one set of reducing gears, and *m* the flange for connection to the working unit. Fig. 40 shows a sectional view of a complete turbine and connections for a single working shaft.

Bearings. The turbine shaft is supported in three bearings. The outer bearing is solid and is held against a ground spherical

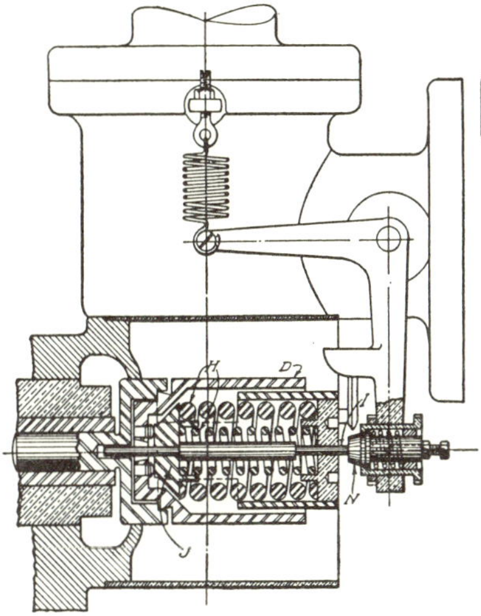

Fig. 41. Details of De Laval Governor and Automatic Safety Vacuum Valve

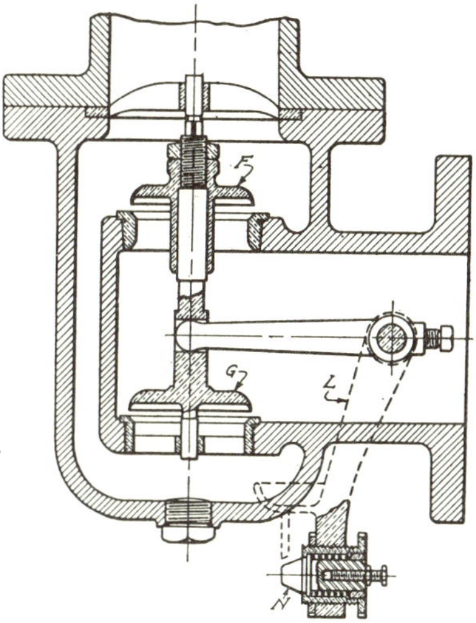

Fig. 42. Details of Bell-Crank Lever and Throttle Valves for Governor Shown in Figure 41

seat by means of a cap and spring. It is made of bronze, lined with babbitt metal. The other two bearings are arranged one on each side of the pinion. They are very long, and hold the pinion accurately in mesh with the gear. They are split to facilitate removal, and have suitable grooves for lubrication. No provision for adjustment is made. The lubrication is supplied from a central reservoir by means of sight-feed lubricators.

Governor. The speed regulation is obtained by means of a simple type of centrifugal governor located, in the geared type of turbine, at the end of one of the slow-speed shafts. It consists of two weights *D*, Fig. 41, hinged on knife-edges, acting on a sliding collar *J*, mounted on a spindle *I*. The governor weights in moving outward push against the collar, moving the spindle outward and at the same time compressing the springs *H*. The spindle in turn presses against the pin *N* at the end of bell-crank lever *L*, Fig. 42. This bell-crank lever operates the throttle valves *G* and *F* shown in Fig. 42.

Riedler=Stumpf Turbine. The first turbine developed by Professors Riedler and Stumpf was of the single-stage type, both pressure and velocity, like the De Laval, but with this radical difference—a wheel about ten times as large in diameter as the De Laval wheel was used and, therefore, the same peripheral speed was obtained with about one-tenth as many revolutions. The reduction which De Laval accomplished by means of gears, Riedler and Stumpf accomplished

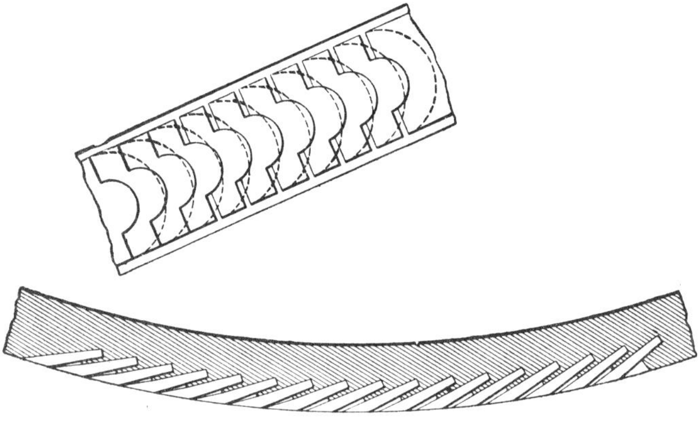

Fig. 43. Milled Buckets in Riedler-Stumpf Turbine

by increasing the diameter of the wheel. By this reduction in the number of revolutions, the error in balance, which, it is claimed, could be brought to less than $\frac{1}{500}$ millimeter, was rendered insignificant.

Their wheels were said to be made of 10 per cent nickel steel with 135,000 pounds tensile strength, and were $6\frac{1}{2}$ to 9 feet in diameter, revolving about 3000 r.p.m. for machines of 2000 to 3000 horsepower. Their single-stage turbine did not meet with general favor, and was usually compounded either by pressure stages or velocity steps, but a description of it will, nevertheless, be valuable.

Instead of using vanes of the De Laval type, U-shaped buckets were milled in the face of the solid wheel, overlapping one another as shown in Fig. 43. The steam jet impinged on the buckets—not on the side of the wheel, as in the De Laval type, but directly upon the face of the wheel—thus permitting a more nearly complete reversal of the steam jet and, other things being equal, a higher efficiency. It will be recollected that if the jet is delivered to the vanes at the side, and at entrance and exit makes an angle with the

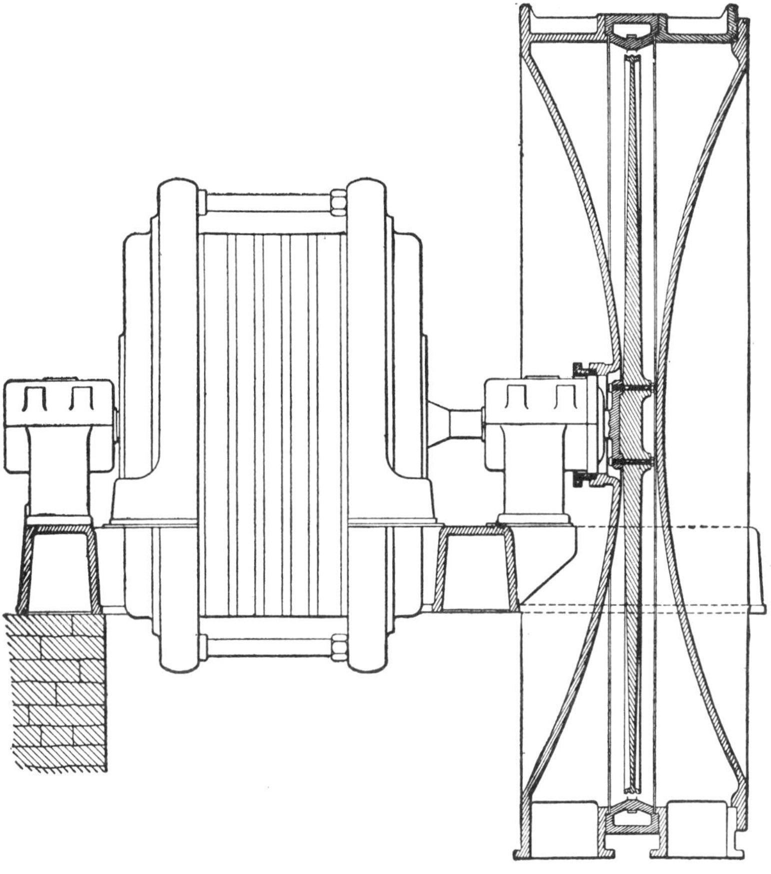

Fig. 44. 20-H. P. Riedler-Stumpf Turbine and Direct-Connected Generator

plane of rotation, the velocity of the jet* is divided into two components. The velocity of whirl, which is equal to $V \cos \alpha$, the angle α usually being 20 to 35 degrees, is the only component that produces a rotative effort.

The nozzles were made with a square instead of an elliptical

*Page 17, Part I.

section at the outlet, and were arranged at regular intervals about the casing, as in the De Laval turbine. With a given size of wheel, the power was increased by increasing the number of these nozzles until steam injection took effect upon the entire periphery of the wheel.

There being only one rotating wheel, it overhung the shaft bearing, thus passing through the casing on one side only, requiring but one stuffing box and, therefore, giving a comparatively small bearing loss. A 20-h.p. turbine of this type, with a direct-connected dynamo, is shown in Fig. 44. Fig. 45 shows details of the wheel. This wheel is fitted with double buckets, which were generally used on the large sizes. A 5000-kw. turbine of this type would require a wheel 20 feet in diameter, admitting steam to the whole periphery and making 1500 revolutions per minute. More details of the Riedler-Stumpf turbines will be described in connection with the compound turbine.

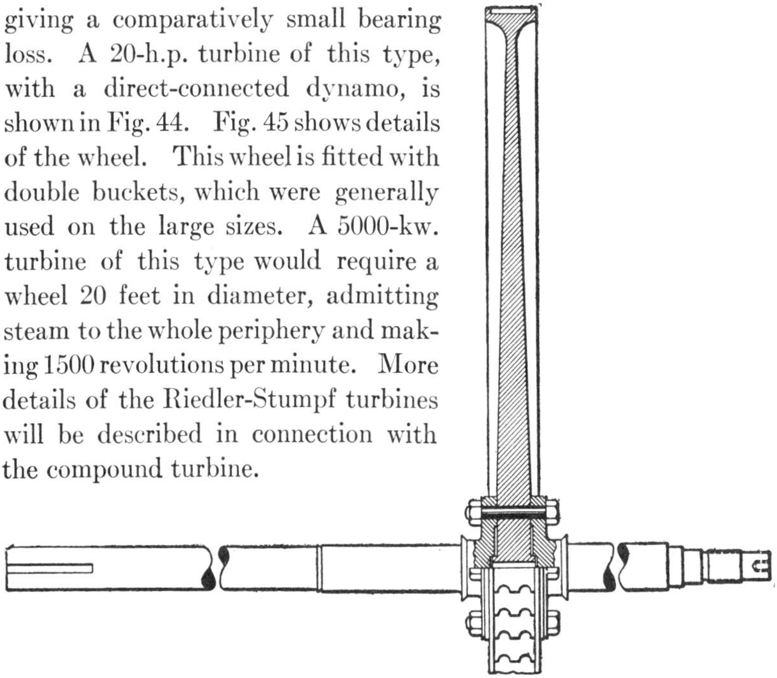

Fig. 45. Detail of Wheel of Riedler-Stumpf Turbine

COMPOUND IMPULSE TURBINES WITH VELOCITY STEPS

It has been shown that steam may be fully expanded to the back pressure in a single nozzle, and the kinetic energy absorbed by passing the jet through several sets of revolving wheels or vanes in succession, each taking out part of the velocity. To employ velocity steps, some sort of reversing device must be arranged to bring the steam back, either onto another bucket of the same wheel or onto a bucket of an adjoining wheel attached to the same shaft. The former method was adopted in the Riedler-Stumpf turbine, the latter in the Curtis. In either case, a simple and compact turbine is the result, but the type has disadvantages already enumerated in "Steam Turbines", Part I, which limit the number of velocity steps that can be economically used to three or four at the outside. Since the work from the fifth action of the steam would theoretically be only one-ninth

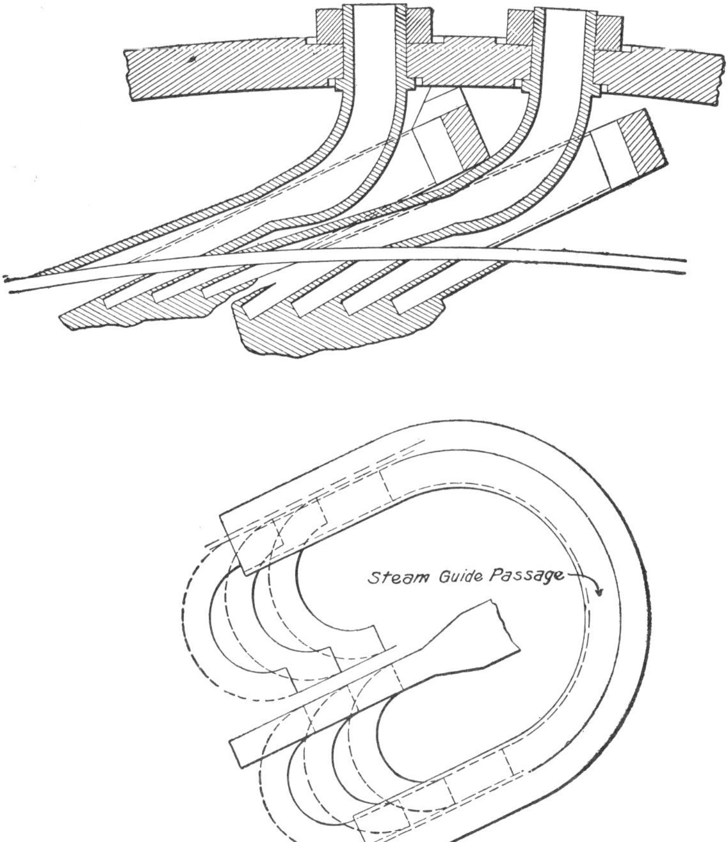

Fig. 46. Double Buckets in Riedler-Stumpf Turbine

of that derived from the first action,* and might easily be consumed in additional friction, it is customary to allow the steam to act no more than four times. Single-stage turbines are not considered practical in sizes above 200 or 300 h.p., it being more economical

*Page 31, Part I.

in such cases to reduce the high steam velocity after the first stage by using at least one other pressure stage.

Curtis Turbine. The earlier forms of Curtis turbine were of the single-pressure-stage type with several velocity steps, and the smaller turbines now made by the General Electric Company are after this pattern. Sizes of 35-kw. and smaller have a single-pressure stage with three velocity steps, that is, three sets of rotating vanes with two intermediate sets of guide vanes. The details of construction are in all ways similar to those of the ordinary form of Curtis turbine, which is compounded both by pressure and velocity, and will be described under the latter heading.

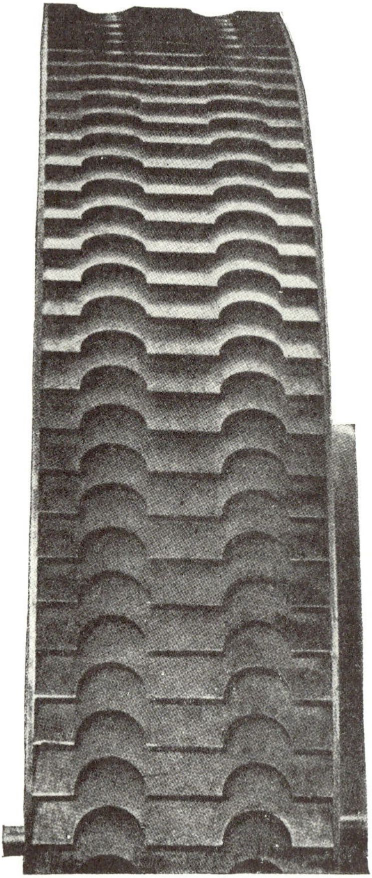
Fig. 47. Double Guide Vanes

Riedler-Stumpf. Large powers of the simple impulse type required either abnormally large wheels or too high speeds of rotation, and it was, therefore, frequently more convenient to extract the velocity from the steam jet in two steps. For powers larger than could be dealt with in the single-stage type, the steam passed successively through buckets of the same wheel, and for still larger powers, pressure stages were employed, as well as the velocity steps. The compound-velocity turbines developed by Professors Riedler and Stumpf had wheels and buckets of the general type described in connection with their simple impulse turbine. The device employed to reverse the direction of the steam and bring it back again to other buckets on the same wheel was described on Page 9, Part I, to which the student is referred.

In one type of their turbine the buckets were double, a small bucket on one side of the wheel being for initial admission; and, since part of the steam velocity was abstracted, it was necessary that, as the steam returned, it should enter a larger bucket which was provided on the other side of the wheel, as shown in Figs. 46 and 47.

Another device of Riedler and Stumpf for reducing speeds of rotation was the employment of counter-running wheels. The guide vanes were buckets cut on a wheel which, instead of being stationary, was free to revolve in a contrary direction. Thus the absolute

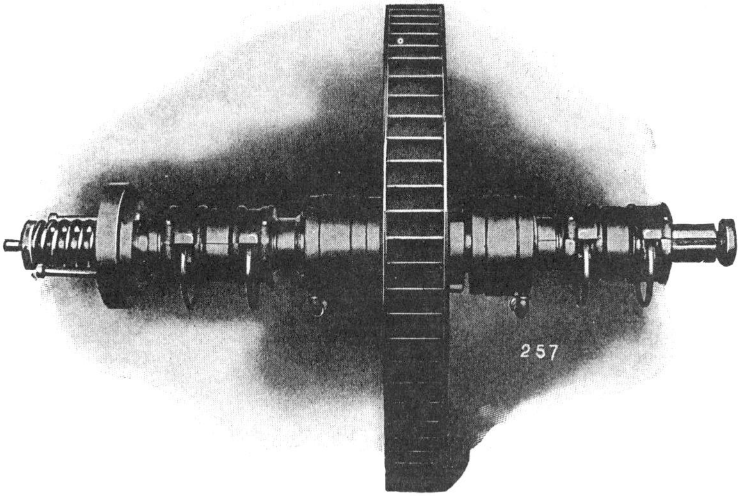

Fig. 48. Rotor and Shaft of Terry Single-Stage Turbine
Courtesy of Terry Steam Turbine Company, Hartford, Connecticut

velocity of each wheel would be half the relative velocity of the two wheels. In a turbine of this type, besides the obvious objection of rotation in two directions, the wheel of initial admission would do more work than the counter-running wheel, because the work absorbed would be in proportion to the difference of the squares of the steam velocities at entrance and exit, and the higher velocities would naturally exist in the first wheel.*

Terry Turbine. The turbine developed and now built by the Terry Steam Turbine Company of Hartford, Connecticut, is, in

*Page 31, Part I.

sizes up to 1000 h. p., of the single- or two-stage, compound-velocity type. The buckets are U-shaped, milled on the face of the wheel, overlapping one another something like the single bucket arrangement of the Riedler-Stumpf machines. Fig. 48 shows the rotor of a

Fig. 49. Terry Horizontal Turbine with Casing Raised for Inspection of Rotor and Bearings
Courtesy of Terry Steam Turbine Company, Hartford, Connecticut

Terry turbine. The steam is expanded in the nozzles to the exhaust pressure or vacuum. As it leaves the nozzles it impinges upon one side of the bucket, reversing through 180 degrees. As it leaves the first bucket, it enters a similar bucket attached to the casing, which reverses its direction through 180 degrees and causes it to impinge again upon another bucket of the wheel, and so on, until the velocity is all absorbed. The reverse buckets are arranged in groups (usually of four), one group for each nozzle, the steam being returned to the wheel as many times as there are reverse buckets in each group. Fig. 49 clearly shows these buckets on the inside of the lifted casing. A crescent-shaped hole may be seen cut in the bottom of each reverse bucket. These holes release a part of the expanded steam and thus reduce the volume in proportion

Fig. 50. Terry Turbine Jet and Reversing Chambers
Courtesy of Terry Steam Turbine Company, Hartford, Connecticut

to the lessened velocity, as otherwise there ought to be successively larger passage areas.

In Fig. 50 is shown how a nozzle section, together with its reversing buckets, is bolted to the steam chamber of the casing. The path of the steam is clearly indicated here.

There are usually four to eight nozzles in the turbine, each being controlled by a hand valve, so that the power may be easily regulated. The main bearings are of the ring-oiling type. As the weight of the rotor is comparatively small and the speed of revolution 1250 r.p.m. for 200 to 300 h.p., large sizes offer no practical difficulty.

Sturtevant Turbine. The B. F. Sturtevant Company of Hyde Park, Massachusetts, builds a turbine in small sizes to drive electric generators and blowers. In sizes of 100 h.p., or less, these turbines have a single-pressure stage, using the steam over and over again on the wheel in much the same manner as is done in the Terry turbine. Powers of 200 h.p., or over, would be built with two or more pressure stages. The wheel is a single forging of open-hearth steel. The buckets, which are the U-shape type, are cut from the solid rim by a milling machine.

The earlier turbines have buckets cut on the side of the wheel, as shown in Fig. 51. Steam entering the outer edge of these buckets passes through the buckets into stationary reverse guides in the casing shown in Fig. 52. At *A* are inserted the nozzles, which are of the ordinary expansion type with elliptical openings. The guides are of two types; about four are U-shape, like the buckets, and return the steam to the wheel again, returning it as many times as there are return buckets; the others, shown at *C*, are cut open at the inner edge in such a manner that the steam, instead of returning to the wheel, is exhausted into the middle of the casing and there allowed to pass out. To avoid a troublesome end thrust in this machine, buckets are cut on both sides of the wheel, thus equalizing the side pressure.

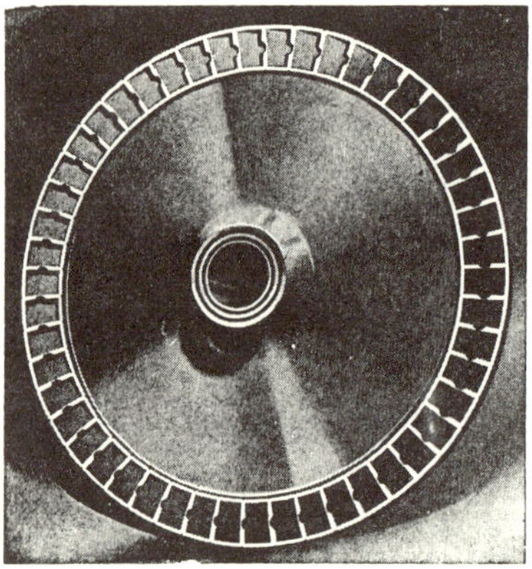

Fig. 51. Wheel of Sturtevant Turbine with Buckets Milled from Side of Wheel

The Sturtevant turbine as built at the present time is illustrated in Fig. 53. The nozzles and reversing buckets are made of Tobin bronze and are cast together in segments. These segments are bolted to the casing, which contains an annular steam chamber. The nozzle is at *A*, with four reversing buckets in front of it and one supplementary bucket behind it, the purpose of which is to utilize any steam which might escape over the back edge of the nozzle. These turbines are built in sizes from 5 to 250 horsepower, the diameters of the rotors ranging from 12 to 36 inches.

Fig. 52. Sturtevant Turbine Showing Reverse Guides in Casing

Fig. 53. Parts of Sturtevant Single-Stage Turbine
Courtesy of B. F. Sturtevant Company, Hyde Park, Massachusetts

De Laval Impulse-Stage Turbine. The class "C" turbines manufactured by the De Laval Company are of the velocity-stage type. They are built with either two or three velocity stages in sizes from 1 to 600 horsepower. Fig. 54 shows a sectional view of a three-stage turbine. The steam enters through a set of nozzles in which it is expanded to condenser or exhaust pressure. The steam then strikes the first impulse wheel. When it leaves this wheel its velocity has been reduced by approximately twice the velocity of

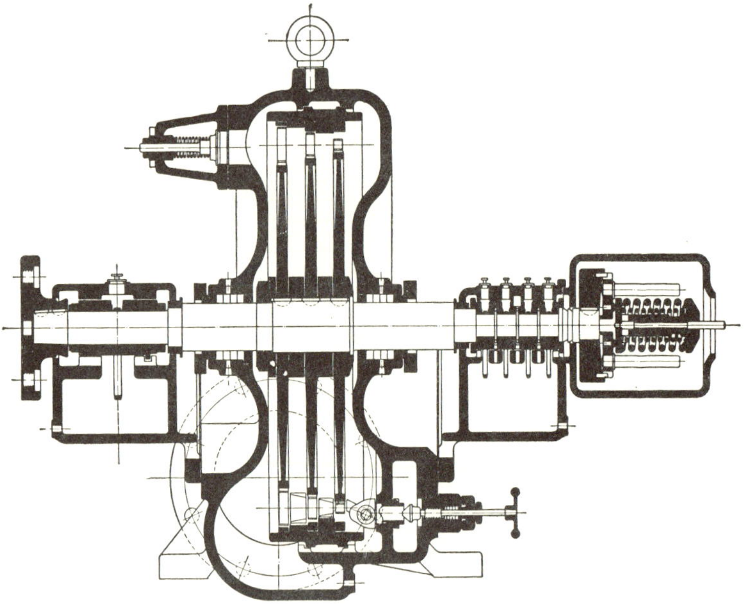

Fig. 54. Axial Section of Class "C" Turbine with Three Velocity Stages
Courtesy of De Laval Steam Turbine Company, Trenton, New Jersey

the vanes. It is redirected by a set of stationary buckets to the second impulse wheel, and so on to the exhaust end of the turbine.

The wheels are forged-steel disks keyed to a heavy shaft which, unlike that in the single-stage impulse turbine, is entirely rigid. The nozzles are of exactly the same type as in the single-stage turbine. Both the moving vanes and the stationary guide vanes have a crescent-shaped cross section and are of practically the same type as is used in the single-stage turbine. The guide vanes are held in steel rings which are split on a horizontal diameter, as shown in Fig. 55.

Westinghouse Impulse Turbine. The Westinghouse Machine Company builds impulse turbines of the re-entry type in sizes up

Fig. 55. Stationary Guide Vanes in De Laval Turbine, and Method of Mounting
Courtesy of De Laval Steam Turbine Company, Trenton, New Jersey

to 500 horsepower. Fig. 56 shows in diagrammatic form the construction of this turbine. The steam is expanded in the nozzle; it then strikes the crescent-shaped vanes of the rotor, giving up a

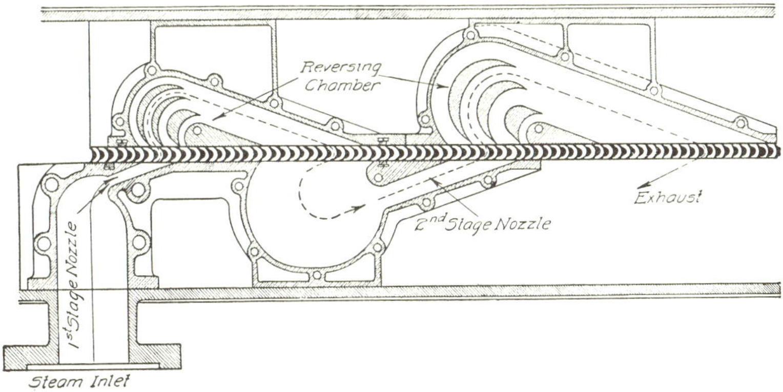

Fig. 56. Section of Westinghouse Re-Entry Type Impulse Turbine
Courtesy of Westinghouse Machine Company, East Pittsburgh, Pennsylvania

portion of its energy. It then enters the reversing chamber on the opposite side of the wheel and is redirected onto the same set of moving vanes, giving up more of its energy. The steam now enters another nozzle, where it is expanded to exhaust pressure, and the operation is repeated. In the small sizes and for pressures less than 125 pounds there is only one nozzle and one reversing chamber.

The construction of the nozzle and the reversing chamber is clearly shown in Fig. 57, while a sectional view of the turbine is shown in Fig. 58. The rotor is a steel disk having a groove in its

Fig. 57. Construction of Nozzle and Reversing Chamber of Westinghouse Re-Entry Type Impulse Turbine
Courtesy of Westinghouse Machine Company, East Pittsburgh, Pennsylvania

periphery. The vanes have a shank at the root fitting into the groove and are riveted to the rotor. The outer end is fitted with shroud ring.

COMPOUND IMPULSE TURBINES, PRESSURE STAGES

In the discussion of compound turbines in Part I it was shown that the available head could be divided into several stages, thus making the steam velocity from stage to stage relatively small, and permitting smaller speeds of revolution. Turbines of this type are, in principle, like a number of De Laval wheels on the same shaft. They consist essentially of a casing which supports a number of diaphragms, dividing the interior into separate cells, in each of which a single impulse wheel containing the vanes is free to revolve. Each stage or element comprises a rotary wheel and a set of nozzles, or distributing vanes, which guide the steam from one chamber to the next and direct it at the proper angle onto the vanes of the wheel in the following chamber. These passages may or may not be of the diverging type, depending upon the drop of pressure from stage to

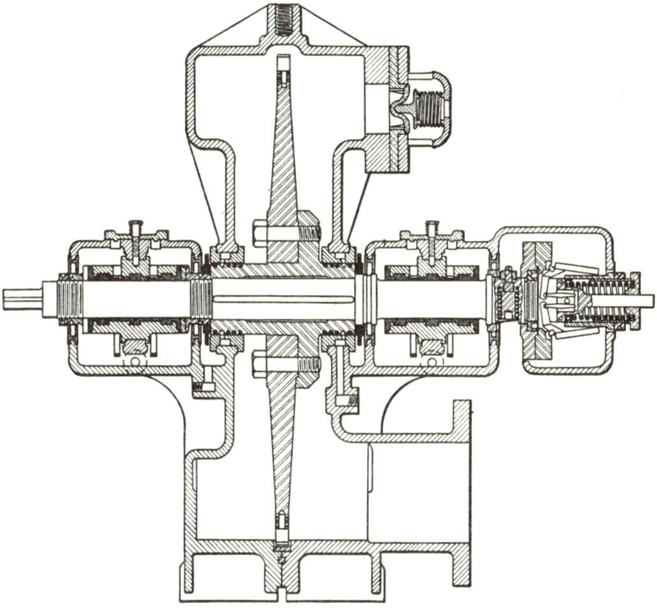

Fig. 58. Sectional View of Westinghouse Re-Entry Type Impulse Turbine
Courtesy of Westinghouse Machine Company, East Pittsburgh, Pennsylvania

stage. In all machines of this type the drop in pressure is so arranged that an equal number of heat units will be given up per stage, which, as will be remembered, does not correspond by any means to an equal drop in pressure.

In such a turbine as this a foreign substance is not likely to injure more than one wheel, for it cannot pass the diaphragm separating the different chambers except through the nozzles and, as there are many stages to such a turbine as this, the machine might run fairly well, even if one or two wheels were removed. It would, of course, give less power and poorer steam economy. The clearance between the nozzles and the vanes should be small to prevent the mingling of the steam jet with the stagnant steam in the casing, but the clearance over the ends of the vanes is of little consequence, especially if a shrouding is used, for there is no tendency for steam to leak by the vanes, the pressure being constant throughout the chamber.

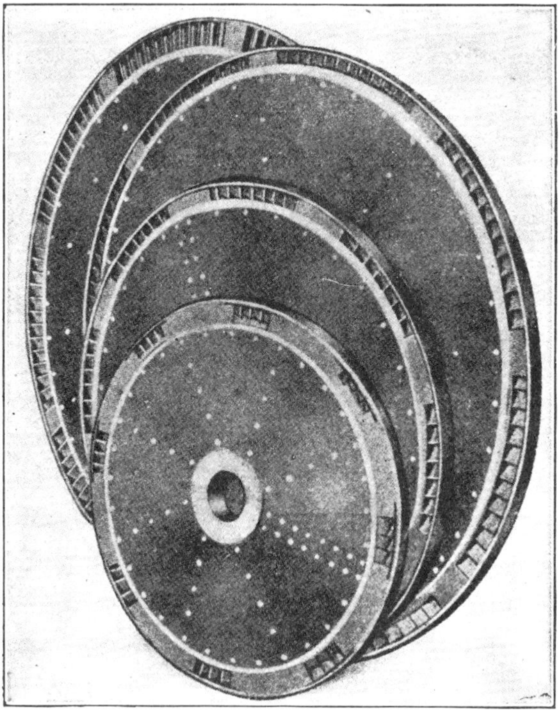

Fig. 59. Group of Diaphragms of Rateau Turbine

Rateau Turbine. The turbine using pressure stages, only, is best exemplified by the Rateau turbine, designed and developed by Professor Rateau, of Paris. His turbine is a horizontal, multi-stage impulse machine and consists of sometimes as many as forty pressure stages, but usually less. The large number of stages employed in these turbines means but little drop in pressure from stage to stage. Hence, the De Laval type of expanding nozzle is not needed, it having been shown in the discussion of nozzles* that, if the final

*Page 19, Part I.

pressure exceeds 58 per cent of the initial pressure, a parallel-sided passage or a slightly converging nozzle is sufficient to permit the proper expansion of the steam and to secure the maximum available energy. In the Rateau turbine, the drop in pressure from stage to stage is much less than the limiting amount mentioned.

Diaphragm. As the steam passes through the turbine, it expands from one stage to the next, and, of course, requires larger passage areas in each succeeding diaphragm. This is accomplished in general by increasing the number of openings rather than by increasing the size of them. The guide passages are arranged in groups, the number in each group being increased and, consequently,

Fig. 60. Lower Half of Turbine Casing Showing Diaphragms in Place, and the Circumferential Grooves for Holding Them

the width of the group widened through successive stages until the openings finally extend entirely around the disk. To provide larger passage areas to take care of still further expansion of the steam, the diameter of the wheel and diaphragm must be increased and, at the same time, the size of the passage openings enlarged. In condensing turbines there are usually three diameters of wheel. In non-condensing turbines, two are generally sufficient, and sometimes only one is used. The nozzles, or distributors, when only a portion of the wheel is open to steam admission, are set to have an angular advance of the preceding group, this advance being proportioned to the speed of the wheels, so that the steam jet as it leaves the revolving vanes will strike the next nozzle directly, avoiding any shock of impact against the solid wall of the casing. Any kinetic energy in the steam as it leaves the revolving vanes of one wheel is, therefore, directly available for use in the next stage. If the steam were brought up sharply against the solid casing wall, this residual energy would be lost to useful work, and a still further loss would result, due to the eddying of the steam in this particular cell. Fig. 59 shows a group of diaphragms of various sizes for this turbine and illustrates the idea of increasing the extent of each succeeding group of nozzles.

There is a distinct advantage in this partial admission at the higher pressures, for, if the admission took place around the entire periphery of the wheel, the height of vane would necessarily be so small that the friction would be excessive. By using partial admission only, the vanes in these stages may be of much greater height than otherwise, and a few high vanes afford the same passage area for the steam that a large number of low vanes would offer, with consequently less friction.

Fig. 61. Two-Piece Stationary Diaphragms with Distributing Vanes

Casing. Circumferential grooves are turned in the inside of the casing to hold the diaphragms in place, as shown in Fig. 60. The larger diaphragms are usually made in two pieces, as shown in Fig. 61. In all of them the shaft passes through collars of antifriction metal with clearances as small as possible, in order to prevent leakage from stage to stage. These collars are frequently provided with a labyrinth packing which will be explained later.

Vanes. The vanes are of crescent shape, similar in cross section to those of the De Laval turbine. The earlier ones had no protection at the outer ends, but the later ones have been provided with a shroud ring to give additional stiffness. The vanes and the shroud ring are generally made of nickel steel, as this is well adapted to resist corrosion.

Wheels. The rotating wheels are usually made of two plate-steel disks. One is flanged at the outer edge to which the vanes are riveted, the other, usually slightly conical, is riveted to the flange of the first disk, and both are riveted to the cast-steel hub. Fig. 62 shows the construction of this type of disk. The conical built-up disk makes a stronger wheel, but a flat disk is sometimes used. Some wheels are turned out of solid steel, increasing in section toward the center to give greater strength. Each wheel is carefully balanced by itself on knife-edges by drilling holes in the disk, the latter clearly showing in Fig. 63. When these wheels are assembled on the shaft there is little likelihood that the complete rotor will be out of balance. Fig. 64 shows the wheels assembled on the shaft.

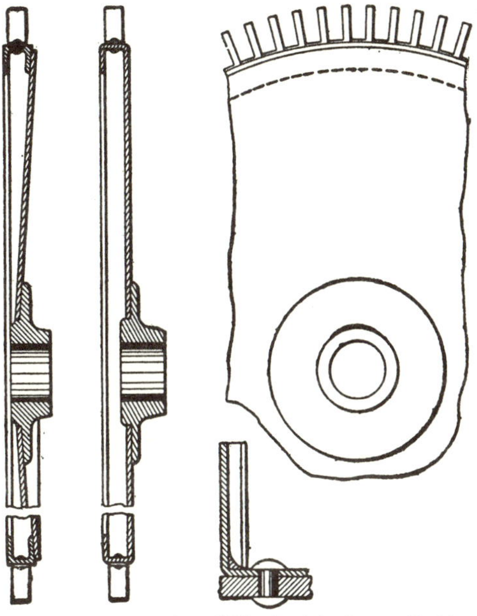

Fig. 62. Construction of Wheels of the Rateau Turbine

By-Pass Valve. This turbine is provided with a by-pass valve to carry overloads. This admits high-pressure steam into the intermediate stages and, although not permitting a complete expansion of such steam, it is an effective means of taking care of large overloads.

Bearings. The bearings are of the plain ring-oiling type, usually provided with water jackets. The shaft not being unduly long, there is little danger of whipping, and as the speeds of rotation are not very high, special precautions are not necessary. Sometimes the turbines are supported by three bearings, as shown in Fig. 65, the high and intermediate stages being separated from the low by a third bearing.

Professor Rateau claims for a 1500-horsepower turbine at 1500 revolutions per minute only $1\frac{1}{2}$ per cent leakage and bearing loss, and a $2\frac{1}{2}$ per cent loss in friction of the wheels against the steam.

Zoelly Turbine. The Zoelly turbine has been developed extensively abroad, and is being manufactured largely through a syndicate of builders including some American firms. It is a turbine

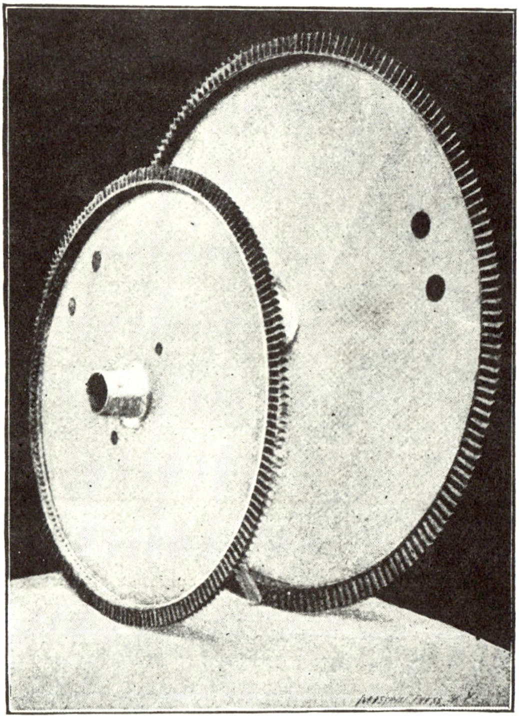

Fig. 63. Wheels of Rateau Turbine, Balanced by Holes in Disk

essentially of the Rateau type, that is, a multi-pressure-stage turbine, but it has fewer stages, usually not over ten for condensing and five for non-condensing, and it differs from the Rateau materially in detail. The blades are very much longer, sometimes being as much as half the diameter of the rotating wheel and, because of the fewer stages, many essential details are different. The turbine is sometimes divided into two parts when built condensing, the high pressure being separated from the low sufficiently for a third bearing to be placed between, as shown in Fig. 66.

Vanes. There being fewer stages than in the Rateau, the steam velocities must be much greater and, consequently, if the rotative speeds are to be the same, the diameter of the wheels must be greater. Exceedingly long vanes are used, which permit of a relatively small wheel disk; hence, the centrifugal stresses in its rim will not be materially greater than in other turbines. Again, the vanes are few in number compared with other turbines and are made as light as possible, tapering at the outer end in order to further reduce the centrifugal force. The expansion from stage to stage is not enough

Fig. 64. Shaft and Wheels of High-Pressure Rateau Turbine

to require diverging expansion nozzles except, perhaps, in the high-pressure end; but the expansion is, of course, very much greater than in the Rateau turbine. In the latter turbine, the roots of the vanes are cut off parallel with the shaft, but in the Zoelly they are cut on a slope, giving a larger outlet than inlet to the vane. It might appear that this is done to permit of steam expansion in the vanes, but that is not so. Expansion is complete in the nozzles just as in the Rateau turbine, but, as there is a much greater expansion from stage to stage, it follows that the area of the steam passages through each successive diaphragm must be greatly increased. This difference is made up by increasing the depth of the openings in the succeeding diaphragm. Therefore, to permit a free passage of steam from an opening having one depth to an opening having a greater

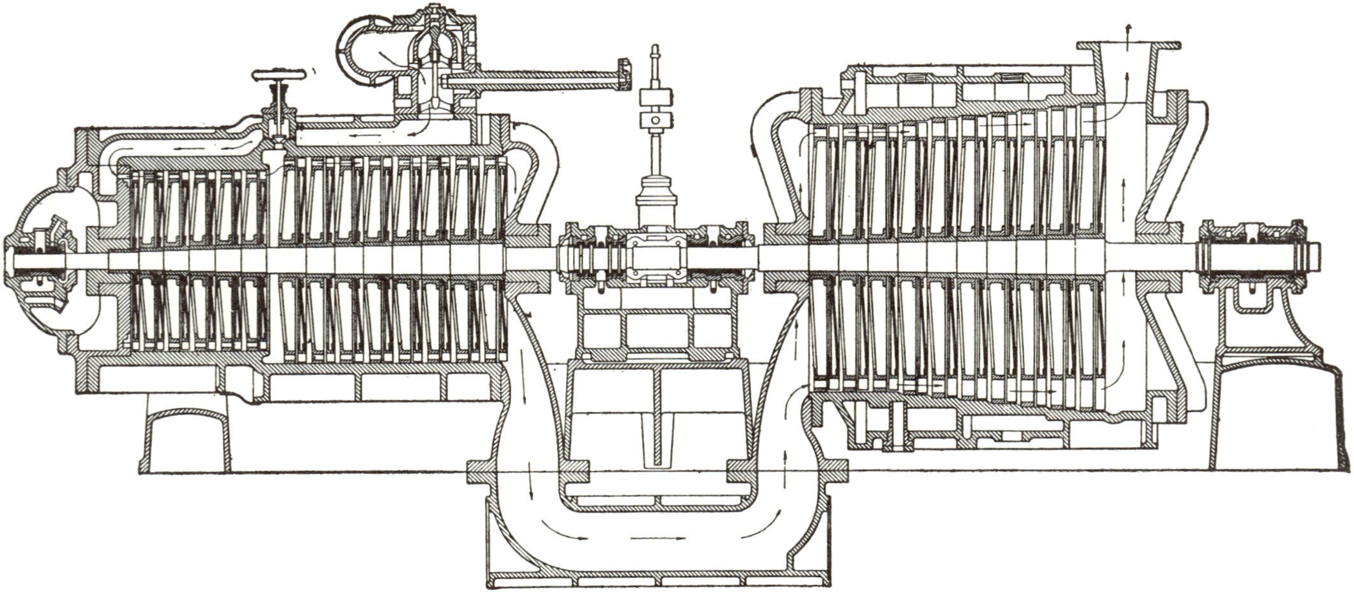

Fig. 65. Three-Bearing Support of Rateau Turbine

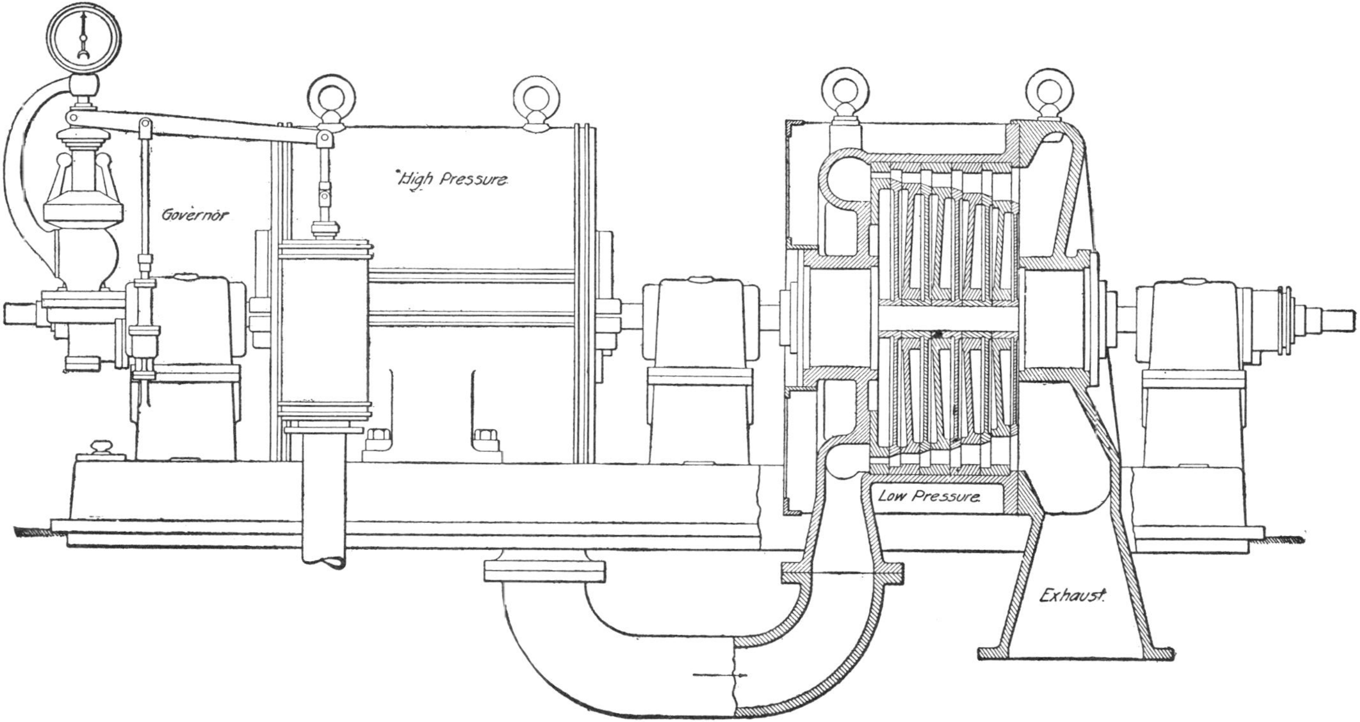

Fig. 66 Zoelly Condensing Turbine, Showing Third Bearing between High and Low Pressure

depth, and to prevent the formation of eddies, it is necessary to slope the root of the vanes. The vanes of the turbines being very much longer than in the Rateau, it is entirely feasible to increase the depth of the steam passages through the diaphragms.

Wheel. The vanes are set in slots cut in the rim of the wheel, and are secured by a clamp ring securely riveted to the main portion of the wheel, as shown in Fig. 67. It is probable that the excessively long vanes produce a considerably greater friction loss, revolving, as they do, in the steam-filled cells of the casing; but as there are comparatively few cells, and therefore comparatively few revolving wheels, it is probable that this friction may not be any greater than would be expected in the Rateau turbine. Economies, as shown by test, do not appear to be essentially different from those of other first-class turbines. Zoelly turbines have been built as shown in Fig. 68. The typical style of long vanes prevails, but in the high-pressure stages the vane is of the double-U-shaped cross section, a detail of which is shown in Fig. 69. The steam jet necessarily impinges tangentially on these vanes instead of from the side.

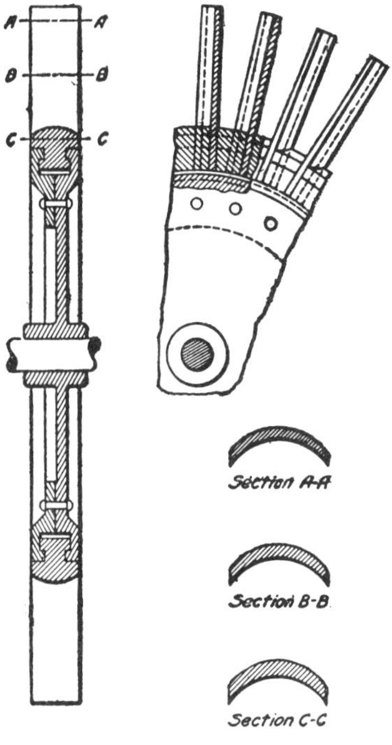

Fig. 67. Detail of Wheel and Vanes of Zoelly Turbine

Hamilton=Holzwarth Turbine. The Hamilton-Holzwarth steam turbine is another turbine of the Rateau type, the chief difference being that, instead of having the admission guides arranged in groups, the admission of steam in the high-pressure end takes place around the entire circumference of the diaphragm. Hence, the wheels of this turbine would be smaller at the high-pressure end than would be the case with the Rateau turbine, and the vanes would be appreciably of less depth. Theoretically, the diameter of each succeeding wheel should increase as the steam expands but, for simplicity of manufacture, it is better to keep a number of wheels

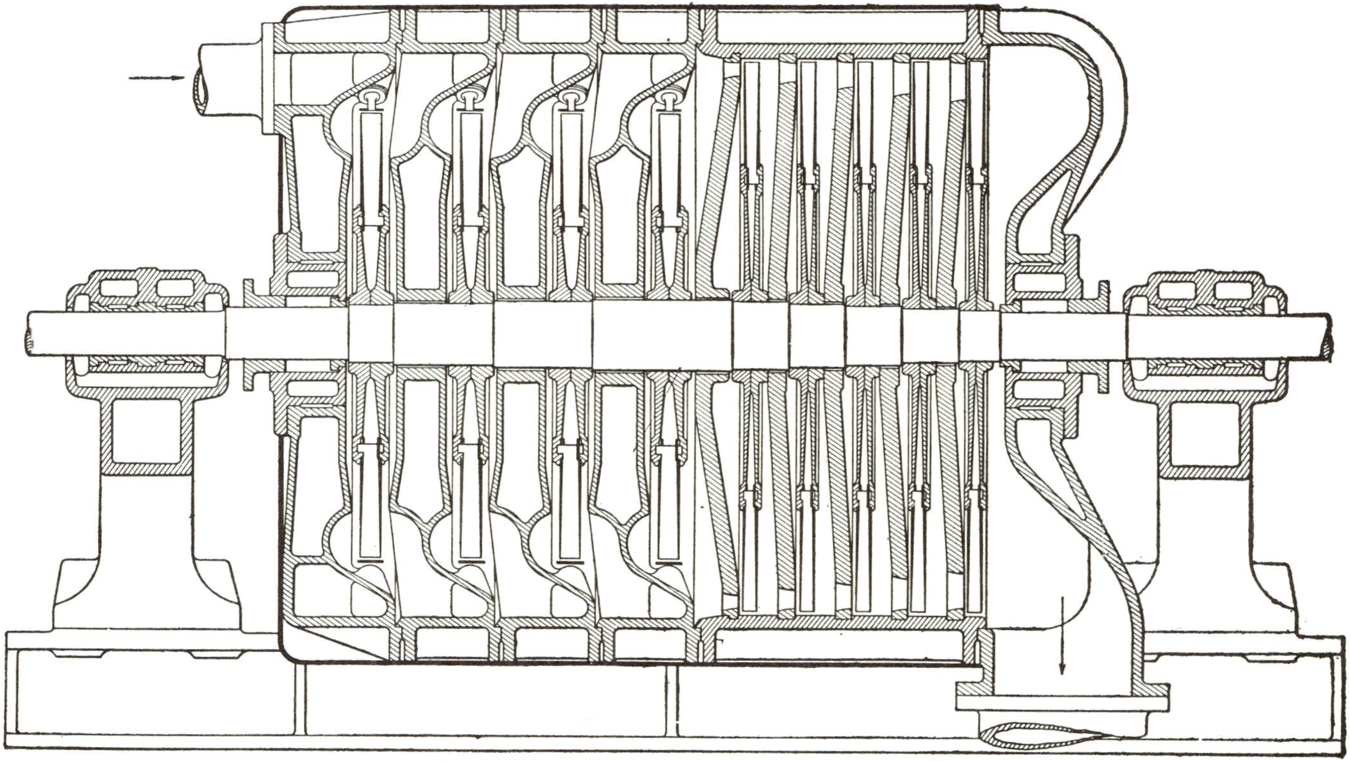

Fig. 68. Section through One Type of Zoelly Turbine

of the same diameter, increasing the length of blades to give larger passage areas. When the point is reached where this is no longer practicable, the diameter of the wheel may be increased considerably and the depth of the blade reduced.

There are approximately the same number of stages in the Hamilton-Holzwarth turbine as in the Rateau, and the speeds of

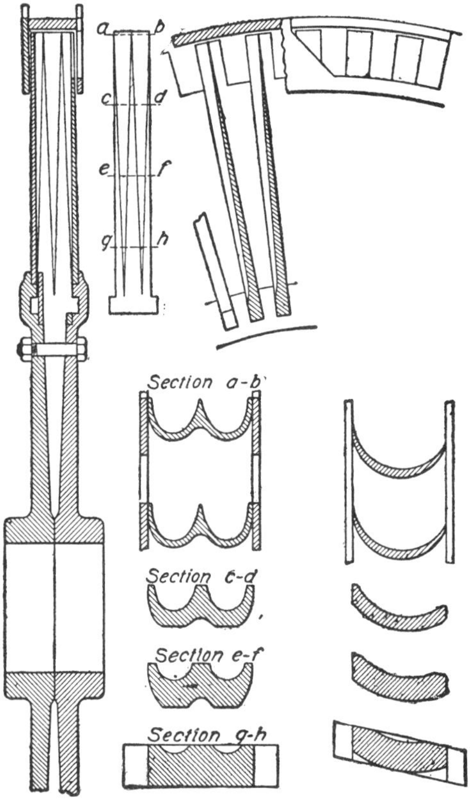

Fig. 69. Details of Construction of the Double-U-Shaped Vanes of the Zoelly Turbine

revolution are not materially different. The running wheel shown in Fig. 70 consists of a steel disk riveted to each side of the steel hub, or spider, which is keyed to the shaft. The outer edges of the disk are flanged outward, leaving a space to take the shank of the vane. The vanes are drop-forged steel of the usual crescent-shaped cross section. The shank of the vane, fitted between two flanged disks, is riveted in place. Since the steam is admitted around the whole circumference of the diaphragm, vanes can be used better than nozzles to give the necessary expansion and direct the steam upon the running wheels. Vanes could be used in the Rateau diaphragm even with partial admission, but in such a case it has seemed simpler to drill openings through the diaphragm.

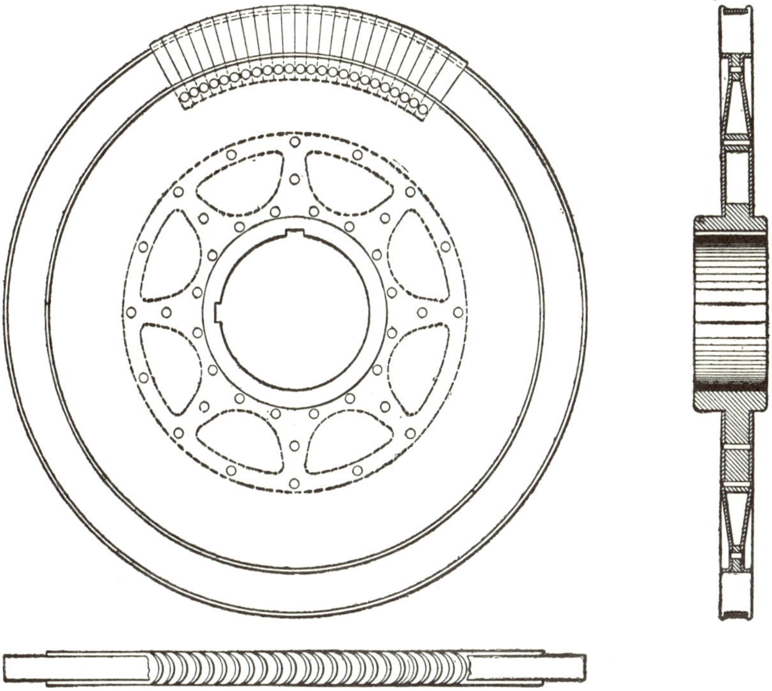

Fig. 70. Details of Wheel of Hamilton-Holzwarth Turbine

For small-pressure drops the vanes may be parallel, top and bottom, but, at the low-pressure end of the turbine where the volumes increase rapidly, they are usually deeper at the outlet than at the inlet, thus forming an easy passage for the steam to the next larger set of vanes.

A diaphragm, Fig. 71, separating the various cells, consists of a cast-iron disk bored loosely to fit the shaft. This disk has a groove *C* into which the shanks of the guide vanes are set, and a rivet holds them in place. A steel band is then shrunk over the ends of these guide vanes, and this band projects into grooves in the casing to hold the diaphragm in place.

The Hamilton-Holzwarth turbine was manufactured by the Hooven, Owens, Rentschler Company of Hamilton, Ohio. It is not being built at the present time.

De Laval Pressure-Stage Turbine. To meet the demand for large-power and slow-speed turbines the De Laval Turbine Company

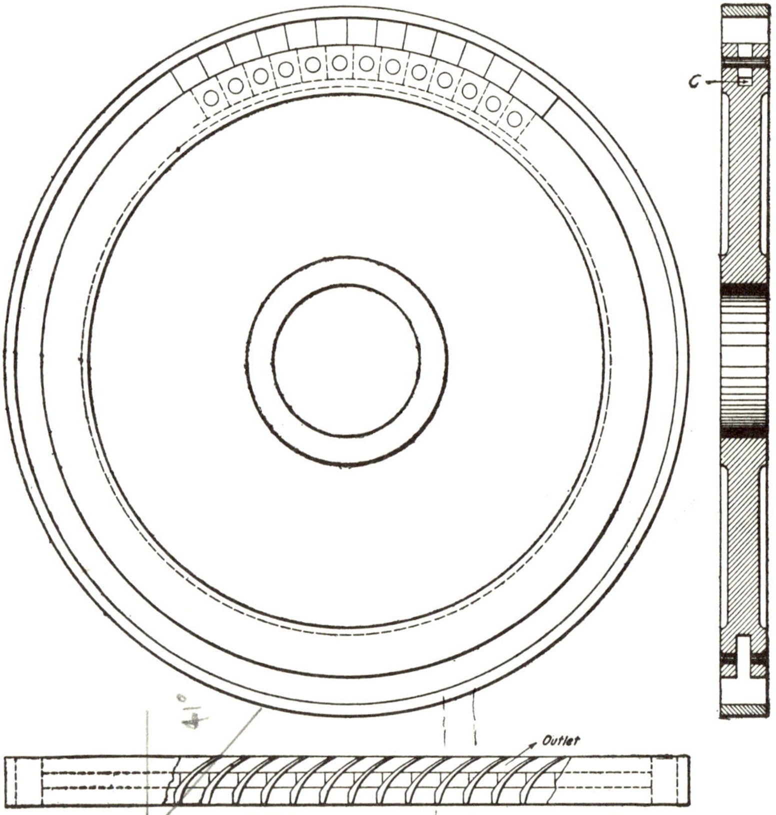

Fig. 71. Details of Diaphragm of Hamilton-Holzwarth Turbine

have recently developed a turbine of the pressure-stage type. The general arrangement is shown in Fig. 72. The rotating member consists of a heavy shaft upon which is mounted a series of disks, each revolving in its own chamber, formed by diaphragms held in the cylindrical casing. The steam is admitted to the first stage through a number of cylindrical nozzles formed in a nozzle ring which separates the steam chest at the right-hand end of the casing from the first wheel.

The vanes are of the same construction as those used in their single-stage impulse turbines, increasing in size progressively toward the exhaust end of the turbine.

With the exception of the nozzles in the first stage, the diaphragms separating the succeeding stages have guide vanes formed around their entire circumference. The vanes are located on the rim of the diaphragm by means of two pins for each vane, and then a

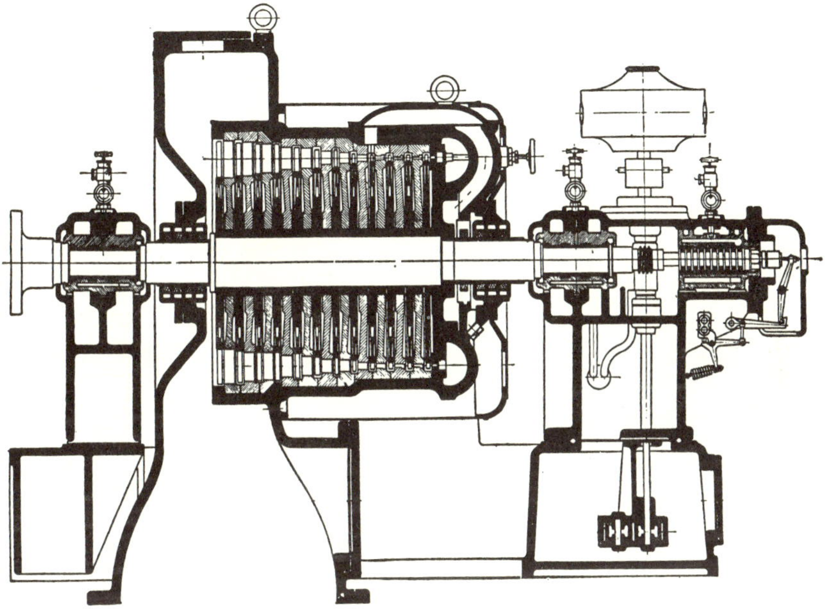

Fig. 72. Axial Section Showing General Arrangement of De Laval Multi-Stage Steam Turbine
Courtesy of De Laval Steam Turbine Company, Trenton, New Jersey

steel band is shrunk over them to hold them in place. These bands project beyond the sides of the vanes and form a strong lining for the cast-iron casing, as shown in Fig. 73.

Wilkinson Turbine. The James Wilkinson Company developed a turbine of the multi-cellular impulse type with comparatively few stages, which is noteworthy because of the packing employed to prevent leakage of steam from stage to stage. It must be borne in mind that in a turbine of the Rateau type there are so many stages that there is comparatively little difference in pressure between one stage and the next, and consequently little tendency for steam to leak through between the shaft and the bushing in the diaphragm. As the number of stages is reduced, the difference in pressure is increased, and it then becomes essential that some sort of packing should be provided; otherwise the leakage may be excessive. In practically all turbines, to prevent leakage of steam from the high-pressure end into the air, a *labyrinth* packing is used, consisting of a series of grooves into which metal spring rings are fitted with slight clearances. These rings are not very different from the spring rings employed to pack the piston of the reciprocating engine, except that there are usually very many more of them. Steam, in order to leak out, must follow a tortuous course between the rings and sides

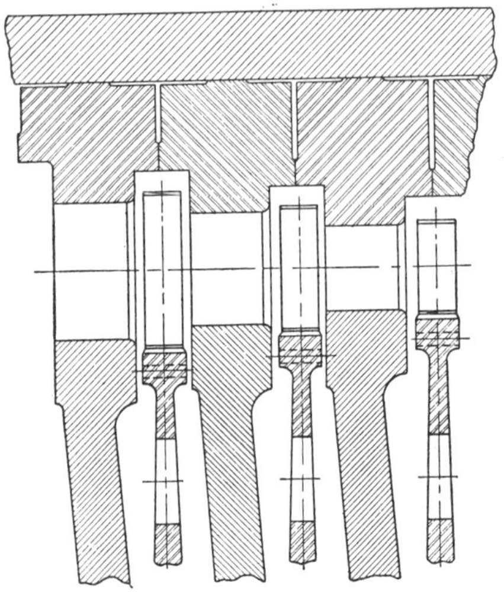

Fig. 73. Partial Section of De Laval Multi-Stage Turbine Showing Steel Retaining Rings

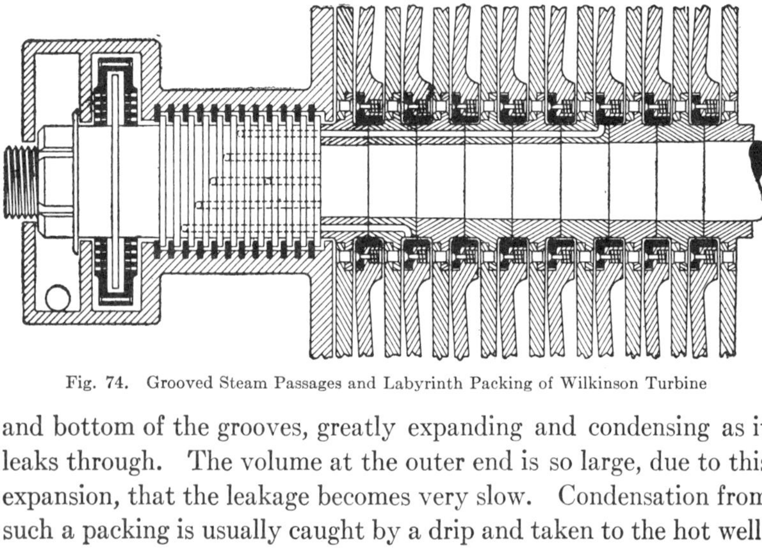

Fig. 74. Grooved Steam Passages and Labyrinth Packing of Wilkinson Turbine

and bottom of the grooves, greatly expanding and condensing as it leaks through. The volume at the outer end is so large, due to this expansion, that the leakage becomes very slow. Condensation from such a packing is usually caught by a drip and taken to the hot well.

The Wilkinson idea is to groove passages along the shaft between certain rings of the labyrinth packing and the diaphragm bushings, as shown in Fig. 74. These grooves permit steam to pass to the diaphragm bushing from a groove in the labyrinth packing, which is at slightly higher pressure than that in the cell at either side of the diaphragm bushing. As a result steam will leak from the labyrinth packing to the diaphragm bushing and, as the pressure of the steam in the cells is less than of that which enters the bushing, this wet steam will expand, thereby vaporizing a portion of its moisture, and will then leak through into the cells of the turbine and have a possible chance of doing some useful work in the lower stages. The leakage from the labyrinth packing will be somewhat augumented by this means, but the leakage from stage to stage will be practically eliminated.

The Wilkinson turbine has a unique governing device but is essentially of the Rateau type of turbine, differing only in detail.

Kerr Turbine. The turbine as originally built by the Kerr Steam Turbine Company of Wellsville, New York, is of the multi-stage type and differs from the Rateau in three important particulars. There are comparatively few stages; the vanes and buckets of the rotating wheels are of the double cup-shape sort, after the style of buckets of the Pelton water wheel; and the steam is directed onto the wheel through nozzles instead of guide passages, striking the wheel tangentially in the plane of rotation instead of at the side. Otherwise the difference between this turbine and others of the multi-stage type is one of detail of construction.

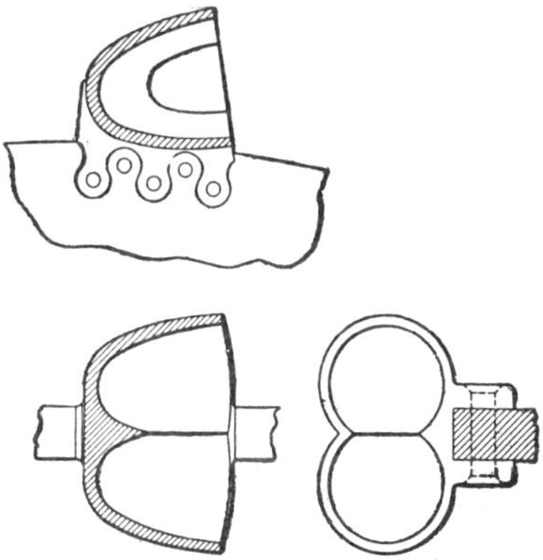

Fig. 75. Buckets of Kerr Turbine

The turbines are built in standard sizes, with wheels 12, 18, 24, and 36 inches in diameter, with one to eight stages, developing up to 600 horsepower. The power developed in a single wheel of given diameter will naturally depend upon the number of nozzles in action.

The buckets are made of drop forgings and are reamed out with a special reamer, each cup being, in cross section, a surface of revolution. Fig. 75 illustrates this type of bucket.

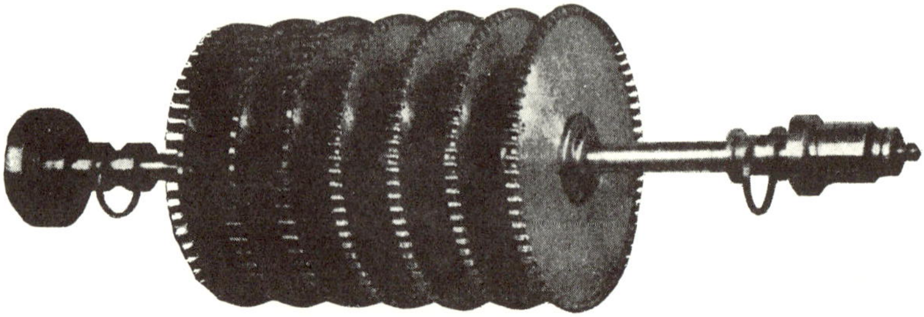

Fig. 76. Kerr Turbine Wheels Assembled on Shaft

The wheel is a solid disk and the buckets may be attached by riveting, as shown in Fig. 75, or by dovetailing, as shown in the same

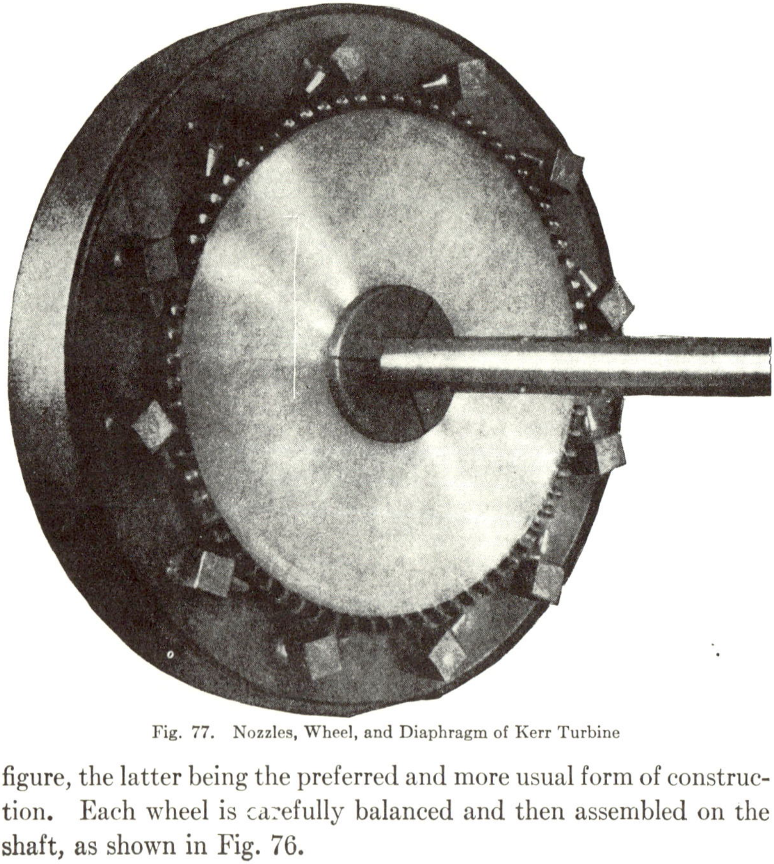

Fig. 77. Nozzles, Wheel, and Diaphragm of Kerr Turbine

figure, the latter being the preferred and more usual form of construction. Each wheel is carefully balanced and then assembled on the shaft, as shown in Fig. 76.

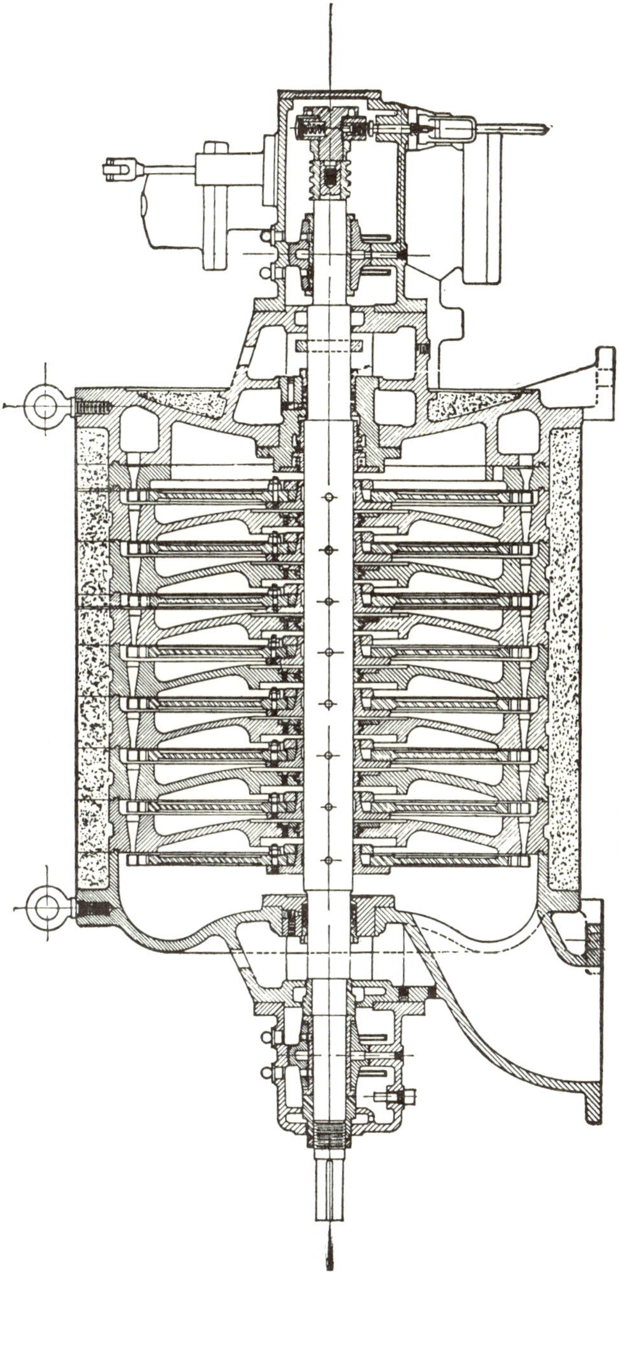

Fig. 78. Section through 8-Stage Economy Turbine
Courtesy of Kerr Turbine Company, Wellsville, New York

Fig. 79. Diaphragm of Economy Turbine, Showing Nozzle Construction. Left Half, Outlet Side; Right Half, Inlet Side
Courtesy of Kerr Turbine Company, Wellsville, New York

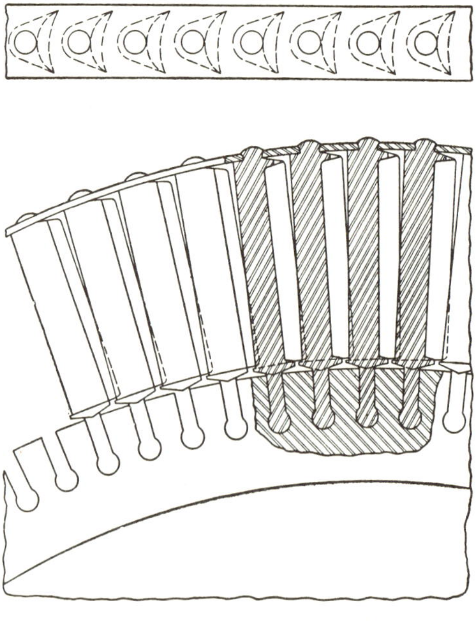

Fig. 80. Section of Bucket Wheel of Economy Turbine, Showing Method of Inserting and Holding Blades

The nozzles are of machine steel, screwed into a steel nozzle body, which is securely riveted to the diaphragm casing. Fig. 77 shows the nozzles, wheel, and diaphragm.

The shell is cast in sections, one for each stage. These are set between two end casings, turned and bored. Tongue-and-groove joints on these castings insure correct alignment, and fibrous packing in the grooves, in addition to metal contact of the surface, insures steam tightness. A casing built in this manner possesses two distinct advantages over a solid casing. There is no probability of cracking due to rapid temperature changes, and the size of turbine may be increased by the addition of more units. The end castings carry the weight of the turbine and have supports bolted to the bed plate.

The shaft, where it passes through the diaphragm, is fitted to a bronze bushing with a few thousandths of an inch clearance. This bushing seats on the metal surface of the diaphragm with latitude for slight side motion. It is kept to its seat by the steam pressure, but can move sideways to accommodate any whipping motion of the shaft. The turbine as built at the present time by the Kerr Company is quite different from the original type. A sectional view of it is shown in Fig. 78. The Pelton form of bucket has been abandoned as well as the separate nozzles. These are formed by walls within the diaphragms and thin metal vanes die-pressed into shape and cast into the diaphragm. They are of slightly converging form. Fig. 79 shows a diaphragm with nozzles.

The buckets as now made are shown in Fig. 80. They are drop forgings held in dovetail slots and riveted. The outer ends are rigidly held in a shroud ring to which they are riveted.

COMPOUND IMPULSE TURBINE WITH PRESSURE STAGES AND VELOCITY STEPS

One of the simplest and most effective ways of compounding turbines is by both pressure stages and velocity steps. The turbine shell is divided by diaphragms into a number of different cells, seldom more then five, except for marine work, where more are necessarily employed. There being comparatively few stages, it will, in most cases, be necessary to employ diverging nozzles so proportioned that the steam may be completely expanded within the confines of these nozzles, from the initial pressure to the pressure in the chamber into which the steam is discharging. It will be remembered that multi-stage turbines of the Rateau type do not require expanding nozzles because of the relatively small drop in pressure from stage to stage.

Turbines of the type now being described differ from the Rateau type in another most important particular. Each cell, or chamber, of this type contains two or more sets of rotating vanes, while turbines of the Rateau type have but one wheel and one set of vanes in each chamber. The steam on leaving the nozzles impinges upon the first set of running vanes, and, as the steam leaves these vanes, it flows into a set of guides of some sort and, as the case may be, is returned to other vanes of the same set, or to a different set of vanes on the same wheel, or to the vanes of a separate wheel. The steam may pass through from one to three sets of redirecting guides, and may

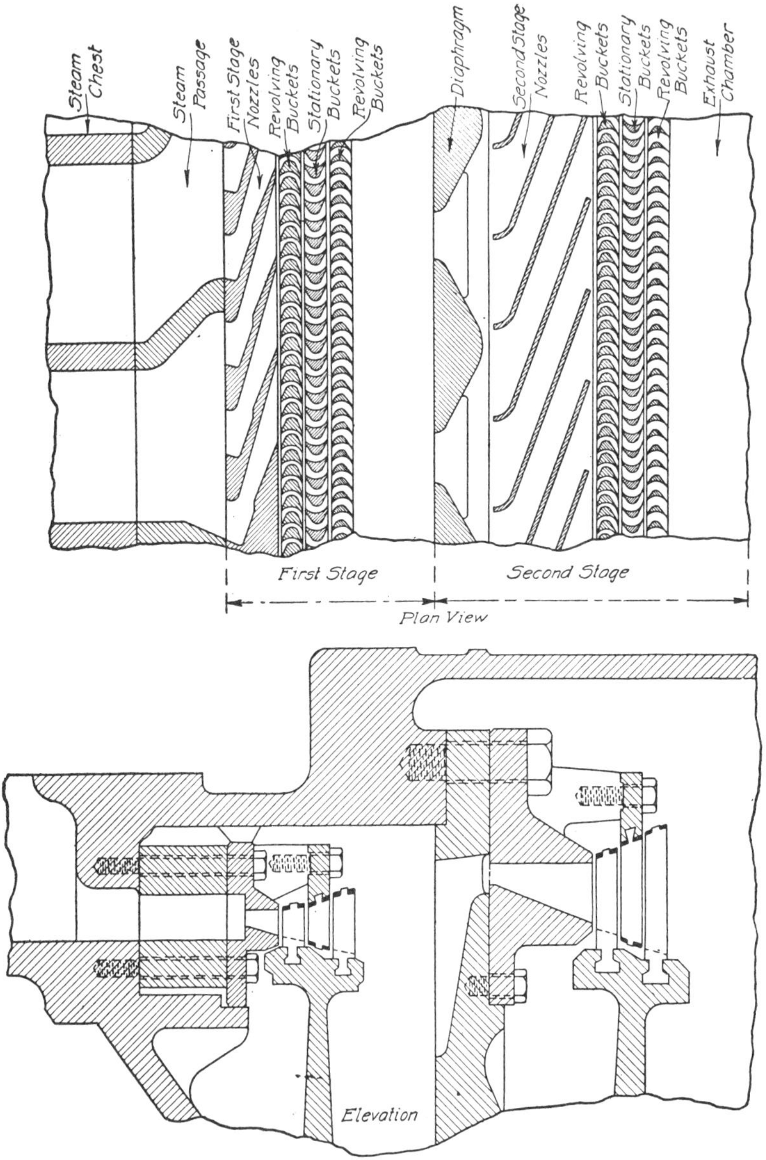

Fig. 81. Plan and Elevation Showing Steam Passages in Curtis Two-Stage Steam Turbine
Courtesy of General Electric Company, Schenectady, New York

impinge upon two to four sets of rotating elements. It is immaterial, so far as the principle of the action is concerned, whether the steam acts successively upon a number of rotating wheels or whether it is returned again and again into different vanes of the same wheel. If the latter form of construction is adopted, the turbine will necessarily be more compact and the rotating shaft will be shorter.

The more velocity steps the turbine has per stage, the fewer number of stages will be necessary, but, in general, it is found more economical to increase the pressure stages than to increase the velocity steps. It must be remembered that in this type of turbine the steam is completely expanded within the nozzle, and that the temperature and pressure of this expanded steam are the same within the confines of any particular cell. As the steam passes through successive rotating vanes, it gradually loses velocity and, consequently, the succeeding passages must be made larger and larger, in order that the same volume of steam may pass through at the lower velocity in a given time. In other words, the passage area must increase in proportion to the reduction in velocity. If this point is not clearly borne in mind when looking at the vanes of a turbine of this type, one might think that the increased size of the passages was to provide for increased expansion of the steam. In the Terry steam turbine, which is of this type, a portion of the steam is allowed to escape as it passes through successive buckets, so that the volume is gradually reduced as the velocity decreases.

Curtis Turbine. Undoubtedly the best-known turbine of this type is the Curtis, which has been developed by the General Electric Company, and is being built at their works at Schenectady, New York, and Lynn, Massachusetts. Rights to build the Curtis turbine for marine propulsion are controlled by the Fore River Ship-Building Company of Quincy, Massachusetts.

For ordinary purposes, up to 9000 kilowatts, the Curtis turbine does not have over four stages, with two velocity wheels per stage, except in the larger sizes of turbine, when a fifth cell is provided, which contains a single rotating wheel without redirecting guides, for abstracting any residual velocity there may be in the steam after passing the fourth stage. The marine turbine of this type installed in the United States Cruiser *Salem* had seven pressure stages with four velocity steps in the first stage and three in each of the others.

Fig. 81 shows in diagrammatic form the principle of the steam action in the Curtis turbine. Steam passes from the steam chest

Fig. 82. 100-Kw. Curtis Steam Turbine Direct-Connected to D. C. Generator
Courtesy of General Electric Company, Schenectady, New York

through the nozzles, each set of which may be closed by a valve operated from the governor. The number open at any one time depends upon the load. There are enough valves and nozzles to take a large overload without the use of the by-pass, which would admit high-pressure steam into the lower stages. Such a by-pass is, however, usually provided. It works automatically and admits high-pressure steam to a set of auxiliary nozzles fitted into the second stage of the turbine, thus increasing the power with only a slight sacrifice in steam expansion and consequent economy.

The nozzles are designed to produce such a drop in pressure from stage to stage that equal amounts of work are done in each stage. This does not correspond to an equal drop in pressure by any means, for there are more heat units in a given drop of pressure in the lower than in the higher ranges. With many stages in a Curtis turbine, the low-pressure diaphragms might be fitted with plain cylindrical nozzles.

Steam enters the first row of rotating vanes* from the nozzles, is deflected from these vanes to the guides, or *intermediates* as they are technically called by the General Electric Company, and is redirected to the next row of moving vanes, and so on, passing from the last row directly into the next and again to the next stage as before.

As has already been mentioned, the number of pressure stages and velocity steps in the Curtis turbine vary with the size of the unit. In sizes from 75 to 300 kilowatts, there are two stages and three velocity steps, as shown in the diagram, Fig. 81. The 500-† to 3000-kilowatt sizes are four-stage, with two velocity steps per stage, while those of over 3000 kilowatts are five-stage, with only a single wheel in the fifth chamber and, of course, no reducing buckets. The turbines, in sizes up to 300 kilowatts, are generally of the horizontal type, the larger sizes being vertical. Fig. 82 shows a 100-kilowatt horizontal turbine, and Fig. 83, a vertical turbine of 9000-kilowatt capacity. A sectional view of the turbine as fitted in the U. S. S. *Salem* is shown in Fig. 84.

It may be interesting to note the speeds of rotation of Curtis turbines of various sizes.

500 kilowatts	approx.	1800 r.p.m.
1000 kilowatts	approx.	1200 r.p.m.
2000 kilowatts	approx.	900 r.p.m.
5000 kilowatts	approx.	800 r.p.m.
9000 kilowatts	approx.	750 r.p.m.
Marine turbines	approx.	250 r.p.m.

*The rotating vanes are called *buckets* by the General Electric Company.

†The General Electric Company builds a special 500-kilowatt vertical turbine having only two stages with, consequently, three rows of revolving vanes per stage.

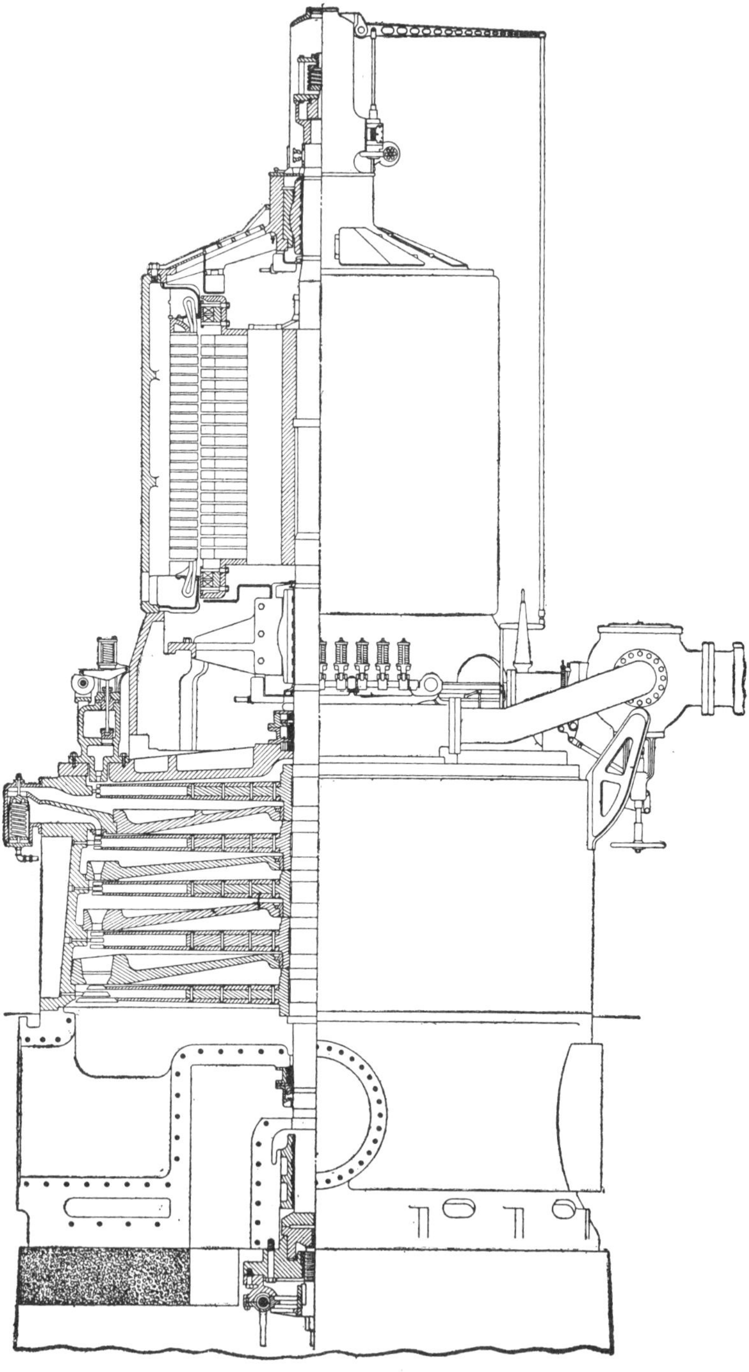

Fig. 83. Vertical Curtis Turbine of 9000-Kw. Capacity

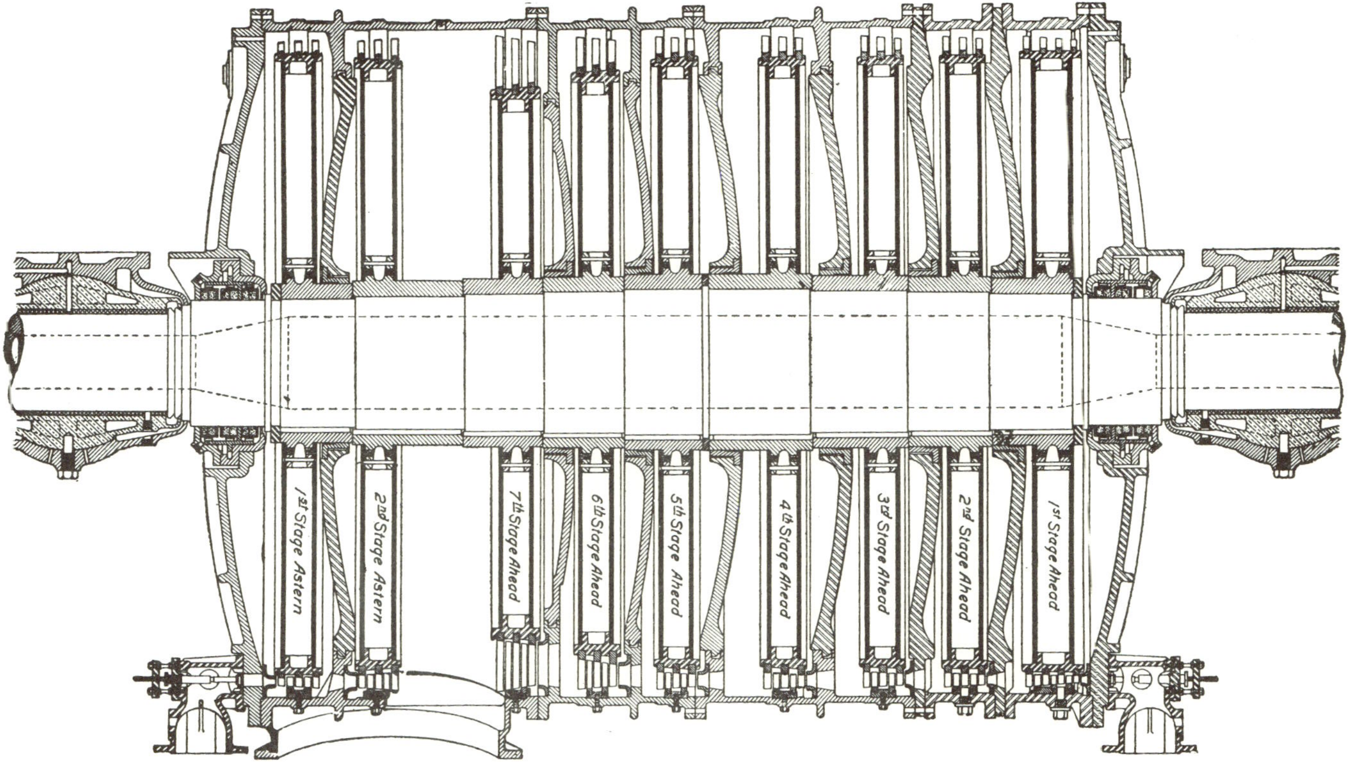

Fig. 84. Sectional View of the Curtis Turbine in the U. S. S. Salem

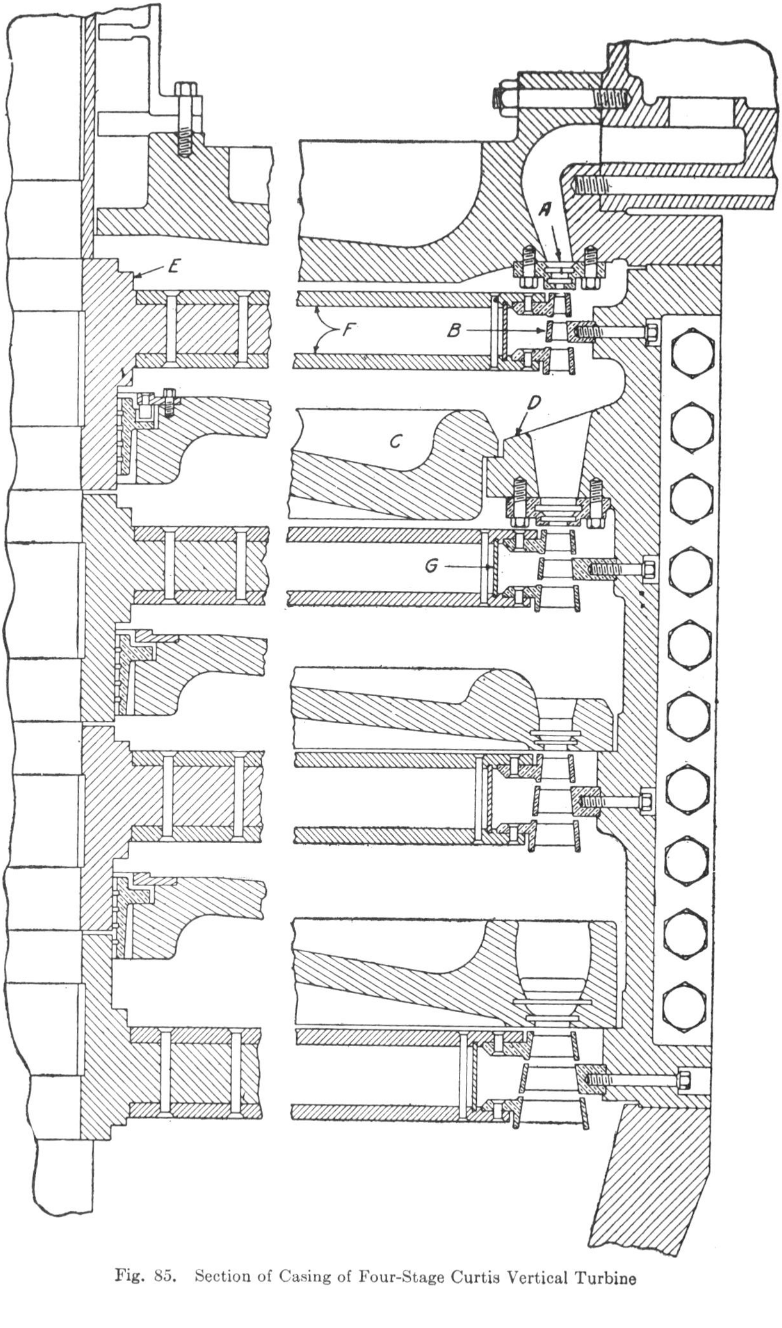

Fig. 85. Section of Casing of Four-Stage Curtis Vertical Turbine

Casing. Fig. 85 shows the section of the casing of a four-stage turbine. It is built of cast iron of four parts for sizes up to 3000-kilowatts, and six parts for 5000 kilowatts and larger sizes. This casing holds the stationary reversing vanes and supports the diaphragm, the details of which are clearly shown in Fig. 85. *A* represents the inlet nozzle for the first stage, *B* the guides or intermediates, *C* the diaphragm separating the first and second stage, *D* the ledge on the casing which supports the diaphragm, *E* the spider of the rotating wheel, *F* the wheel plates, and *G* the distance piece at the outer rim of the wheels.

Fig. 86. Group of Nozzles for High-Pressure End

Nozzles. The nozzles are grouped together, not as in the Rateau turbine, but in one single group for each stage, thus admitting a single steam belt to a part of the wheel periphery only. The Rateau turbine also admits steam to only a portion of the periphery, but in this turbine there are several groups of nozzles instead of one, arranged at equal intervals around the periphery.

Fig. 87. Portion of Diaphragm, Showing Construction of Buckets

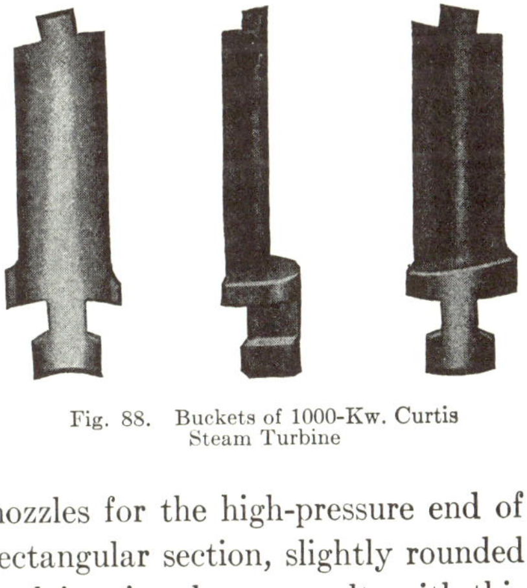

Fig. 88. Buckets of 1000-Kw. Curtis Steam Turbine

Fig. 86 shows a group of nozzles for the high-pressure end of the turbine. The outlet is of rectangular section, slightly rounded at the corners, which the makers claim gives better results with this type of turbine than the elliptical outlet used in the De Laval. The nozzles are reamed out of bronze castings and riveted to the casing.

Vanes. The vanes, usually called buckets by the Curtis manufacturers, are crescent-shaped in cross section, as is common for impulse turbines.

The construction of buckets and wheels is shown by Figs. 87 and 88. The wheel is a steel disk with the rim enlarged so that a groove of dovetail section may be cut in its periphery. The buckets have a corresponding dovetailed root which fits snugly into the groove of the wheel. At intervals the groove has openings for the insertion of buckets, Fig. 87. These openings are then closed by means

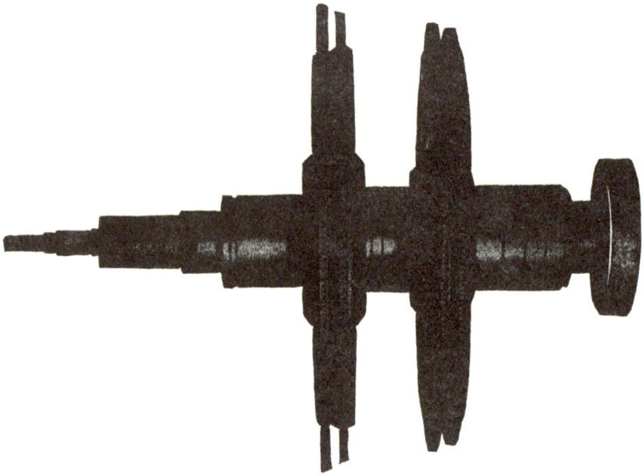

Fig. 89. Rotating Wheels of the Curtis Steam Turbine, Showing Two Sets of Vanes

of a spacing block. After the buckets are assembled, the shroud ring is riveted to their outer ends by means of the small projections shown on the buckets. This ring serves the twofold function of reducing the vibration of the buckets and preventing the jet of steam from spilling over their ends.

Wheels. Fig. 89 shows the rotating wheels of a Curtis turbine, with two sets of vanes so spaced that, as the wheel rotates, there will be one set of vanes on each side of the guide vanes, or intermediates. This is equivalent to having two rotating wheels, but in construction is much simpler. Each wheel, as shown, rotates in a chamber by itself. Figs. 89 and 90 show the old style of bucket segments which offer a very rough surface. In the present style, Fig. 87, the wheels have no external webs and no riveting. The wheels, except in the larger sizes, are made of solid steel, securely keyed to the shaft at the hub. This hub varies in length according to the stage in which it revolves, the low-pressure stages necessarily being wider than the high-pressure stages. The wheel gradually tapers toward its periphery, thus maintaining a section of approximately uniform strength. Wheels for the largest sizes of turbine are built up of a cast-steel hub, or spider, to which are riveted steel disks, one on each side, and between them is riveted a cast-steel distance ring. The vanes are riveted in segments onto the outer edges of the disks, as is clearly shown in Figs. 83 and 84.

Guides. The guide vanes, or intermediates, are necessarily in one group of sufficient extent to catch the steam belt as it issues from the rotating wheels. The group naturally extends at first over only a small arc of the circumference, but this arc increases in extent in each stage until both nozzles and intermediates entirely surround wheel and casing. The guide vanes are in cross section like the revolving vanes, but are set in a reverse position. They are set with a certain angular advance from the nozzles, depending upon the speed of rotation, so that the steam leaving the vanes will strike full upon them. Like the vanes, they have to provide a successively increasing passage area to allow for the lower velocity of the steam, if more than one set of guides is used per stage.

Fig. 90. Revolving Buckets for Curtis Steam Turbine
Courtesy of General Electric Company

Diaphragm. The style of diaphragm is best shown in Figs. 84 and 85, which also clearly show the construction of the large wheels. The diaphragm is an iron casting—steel in the higher stages because the drop in pressure is greatest there, and, consequently, a greater load on the diaphragm—slightly dished in shape toward the high-pressure side to give greater strength. It is provided with bronze bushings where the shaft passes through, these bushings being fitted with slight clearance to prevent leakage. The ends of the casing are packed with carbon packing to prevent leakage of steam into the air at the high-pressure end, and leakage of air into the turbine at the low-pressure end, where leakage would have a detrimental effect upon the vacuum.

Bearings. All sizes of the Curtis turbine of 500 kilowatts and over are of the vertical type with only one working bearing, located at the lower end of the shaft*. This bearing consists of a short cylindrical block of cast iron fitted with two dowel pins and a key, as shown in Fig. 91, and a corresponding cast-iron block having a hole through its center, into which a pipe is threaded for supplying some form of lubricant, either water or oil.

Fig. 92 shows the bearing assembled on the lower end of the shaft. The upper bearing with dowel pins and key fits into corresponding dowel holes and keyway in the bottom of the shaft, and

Fig. 91. Bearing Surfaces in Step Bearing of Curtis Turbine
Courtesy of General Electric Company, Schenectady, New York

rotates with it. When the oil is supplied to the bearing, which is, of course, under a high pressure, it fills the central circular space between the blocks and forces them slightly apart. The oil then escapes between the annular edges of these two blocks and is collected into a drain and returned to the original supply. If water is used for a lubricant, it is allowed to flow up into the base of the turbine and mingle with the exhaust steam on its way to the condenser. The pressures maintained by the lubricating pump in practice vary from 180 to 450 pounds per square inch. It is thus seen that the two bearing blocks do not come into actual contact, but that the weight of the turbine is supported upon a film of lubricant. Should the lubricating pump fail in its supply, no more serious damage would

*The horizontal construction is favored at the present time and the General Electric Company build all sizes of turbines in that style.

occur than the abrasion of the step bearing, and a new one could readily be inserted, as the figure will show.

Riedler-Stumpf Turbine. The turbine developed by Professors Riedler and Stumpf for the larger powers necessitating lower speeds was provided with two to four pressure stages, with two velocity steps per stage, and this turbine expanded the steam on the same principle as the Curtis, but the details of construction and general

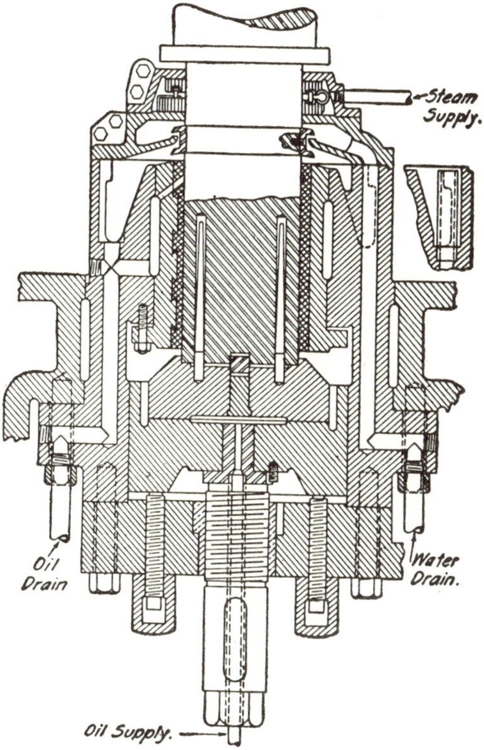

Fig. 92. Step Bearing of Curtis Turbine

arrangement were entirely different. In the two-stage, and even in the four-stage, turbine the overhung type of wheel developed in the single-stage turbine was adhered to, the generator being between the two turbine wheels with a bearing at each end. Fig. 93 shows this arrangement in a 5000-kilowatt, two-stage turbine revolving at 750 r.p.m., with two velocity steps to each stage. In this particular turbine, the steam is returned through U-shaped guide passages to a second set of buckets on the same wheel, these buckets being larger than the first because of the lower velocity. Fig. 94 shows a 500-kilowatt, four-stage turbine of the same type. These have been built both vertical and horizontal, the vertical arrangement resembling externally the Curtis turbine.

Fig. 95 shows a vertical four-stage, two-step turbine of 750 r.p.m. developing the same power as that shown in Fig. 94.

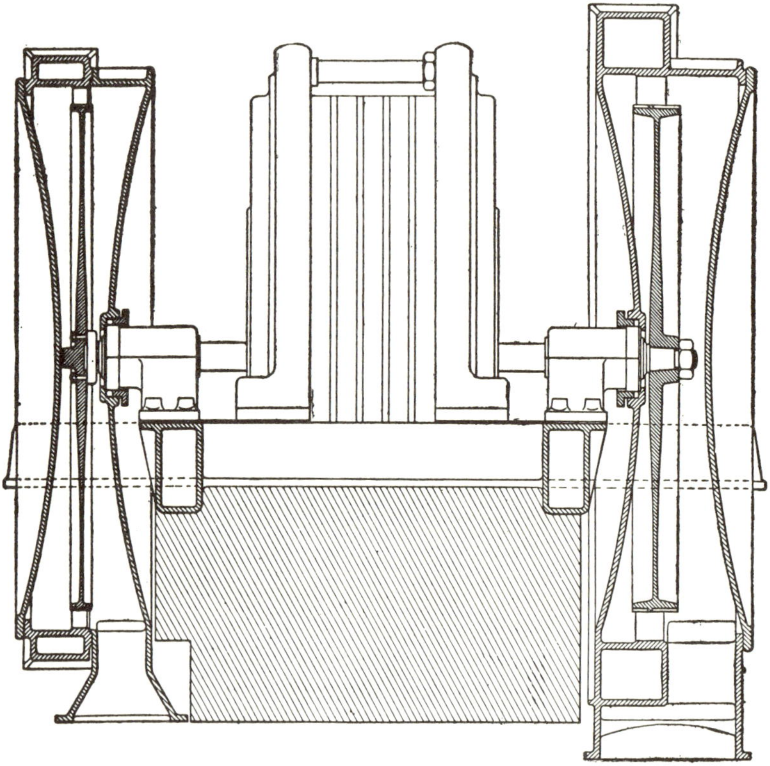

Fig. 93. Overhung Wheels of the Riedler-Stumpf Turbine

The Reidler-Stumpf turbine was formerly manufactured by the *Allgemeine Elektricitats Gesellschaft* of Berlin, but, as this company is now licensed to manufacture under Curtis patents, it does not appear that the manufacture of the former type is being actively carried on at the present time.

Terry Turbine. In the medium sizes of turbines, up to about 360 kilowatts, which are to operate condensing, the Terry steam

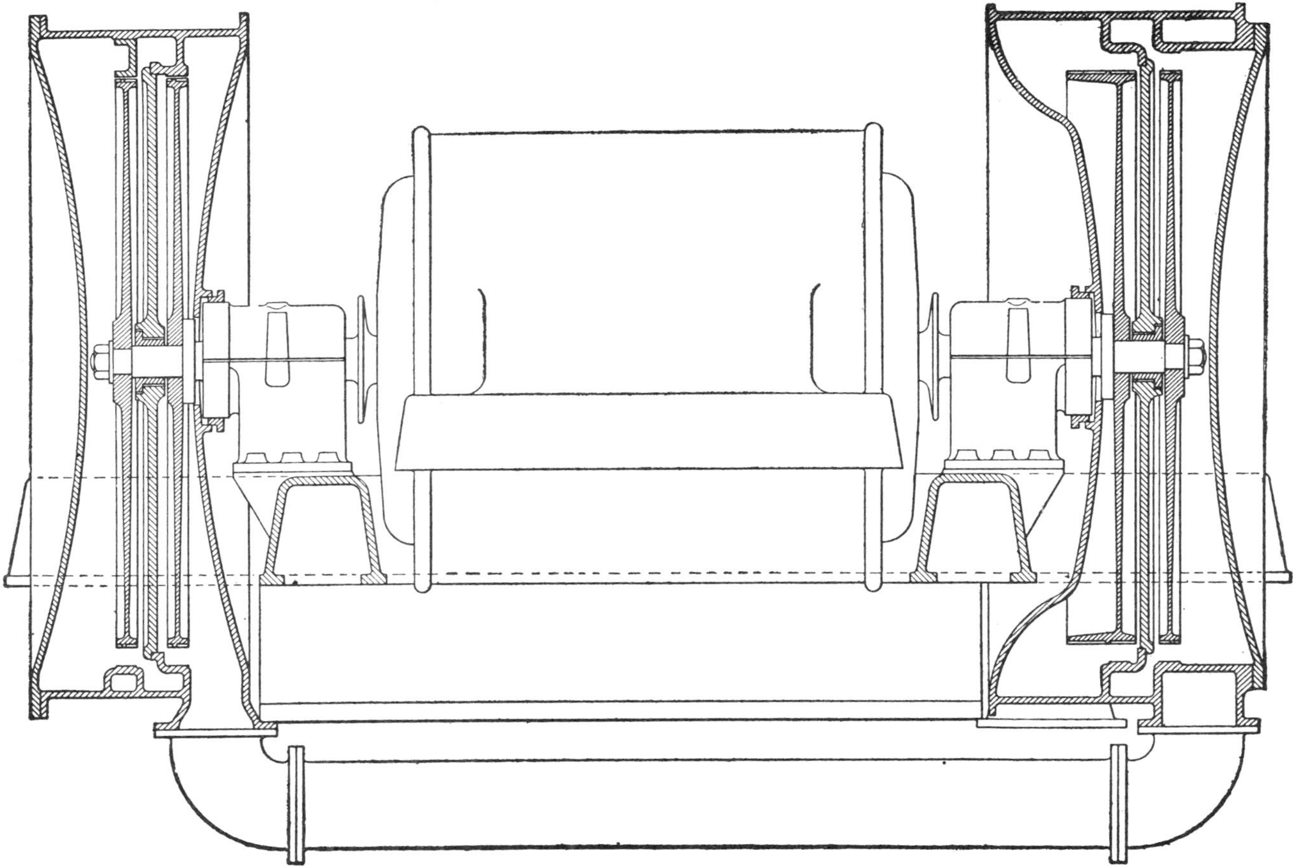

Fig. 94. 500-Kw. Four-Stage Riedler-Stumpf Turbine

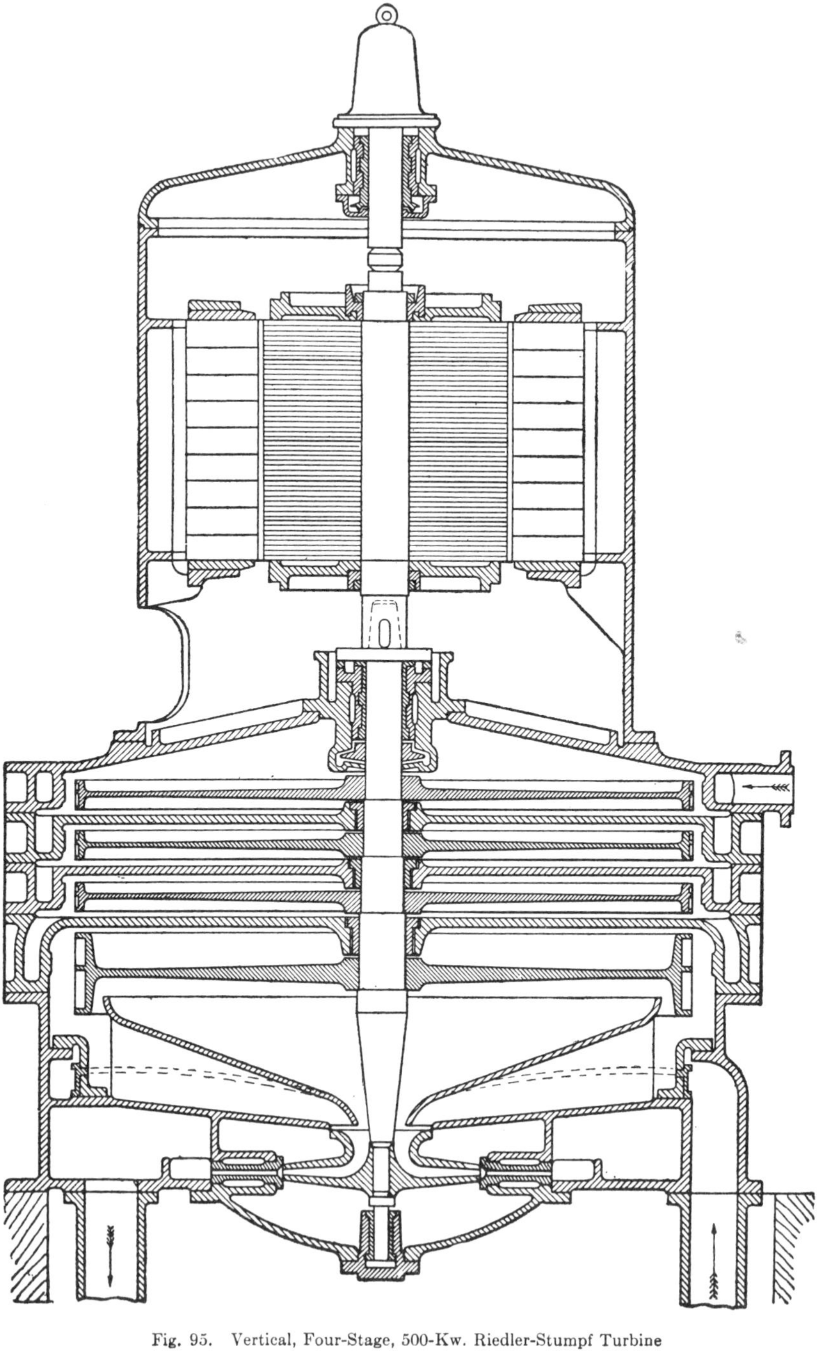

Fig. 95. Vertical, Four-Stage, 500-Kw. Riedler-Stumpf Turbine

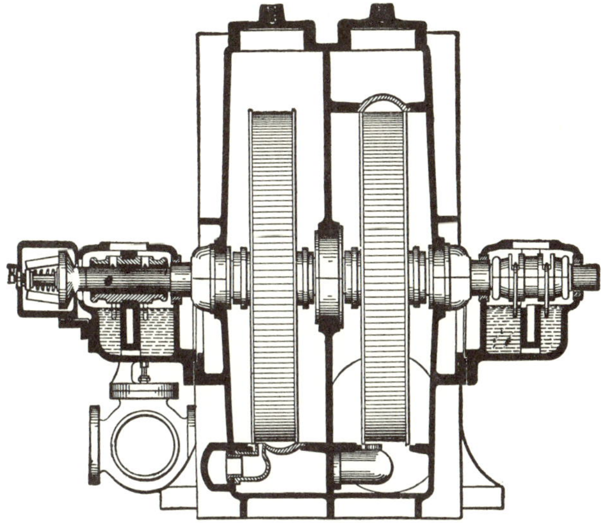

Fig. 96. Section Through Two-Stage, Condensing Terry Steam Turbine

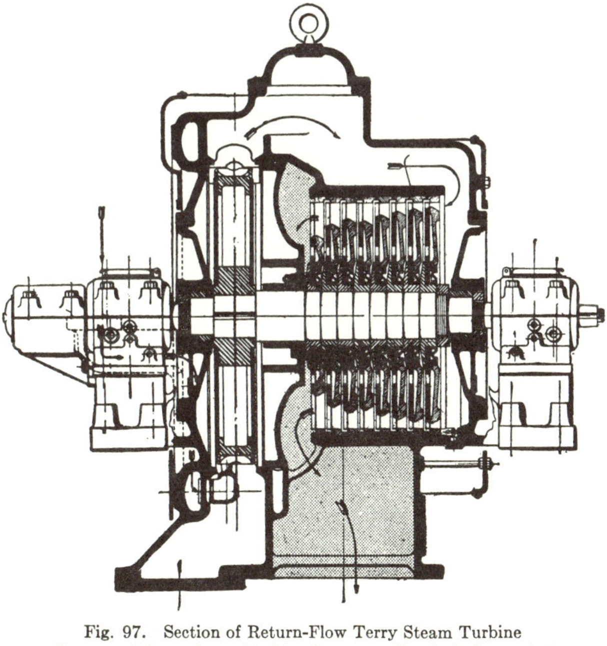

Fig. 97. Section of Return-Flow Terry Steam Turbine
Courtesy of Terry Steam Turbine Company, Hartford, Connecticut

turbines are of the two-stage type, as illustrated in Fig. 96. The turbine consists essentially of two single-stage wheels in the same casing, separated by a diaphragm. The nozzles of the first stage expand the steam to about atmospheric pressure and, after passing through the wheel, it is further expended to condenser pressure in the nozzles of the second stage.

Fig. 98. Return-Flow Steam Turbine with Cover Lifted
Courtesy of Terry Steam Turbine Company, Hartford, Connecticut

More recently the Terry Steam Turbine Company have developed the turbine shown in Fig. 97. This, as will be seen, consists of a Terry turbine wheel in combination with a Rateau pressure stage element, the Terry wheel being at the high-pressure end of the machine. An interesting feature of this turbine is the so-called return-flow principle adopted in its construction. The steam, after having passed through the Terry wheel, is at slightly above atmospheric pressure. It now travels to the other end of the turbine and passes through the Rateau element in the reverse direction. The

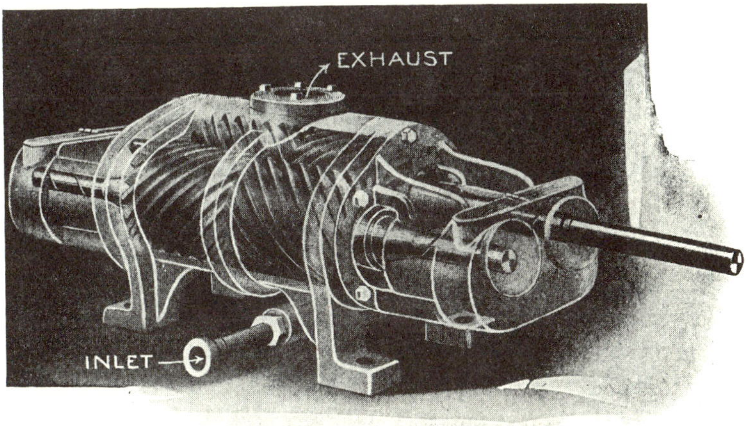

Fig. 99. Phantom View of Spiro Turbine, Showing Extreme Simplicity and Compactness
Courtesy of Buffalo Forge Company, Buffalo, New York

purpose of this is to prevent leakage of air into the condenser where the shaft passes through the casing. Fig. 98 shows an external view of this turbine with cover lifted.

Buffalo Forge Spiro Turbine. The Buffalo Forge Company have developed a novel construction of turbine called the Spiro turbine, a phantom view of which is shown in Fig. 99. As will be seen, it consists essentially of two double helical gears rotating in a cast-iron casing, which closely surrounds them. The steam enters through two openings, one on each side of a central ridge in the bottom of the casing, Fig. 100, and is admitted to a tooth space at the center of the rotors. As the gears rotate, this tooth space extends in length across the entire width of turbine, thus expand-

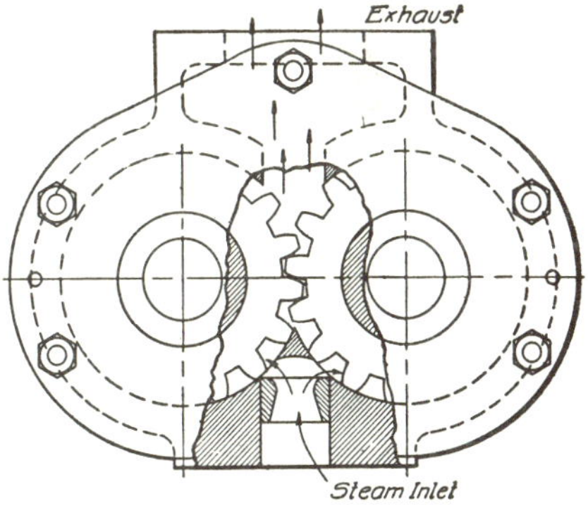

Fig. 100. Section of Spiro Cylinder and Rotors at Midlength

REACTION TURBINES

ing the steam. The maximum ratio of expansion is stated by the makers to be 1 to 6. This turbine, therefore, is only run non-condensing and is useful where steam economy is of secondary importance compared with simplicity and compactness.

REACTION TURBINES

In this type of turbine, the steam expansion takes place, not alone in the nozzles and guide passages, as in all the types of turbines previously described, but in the revolving vanes as well, approximately half the expansion taking place in each. In this type of turbine the difference in pressure on the two sides of the vanes brings about a leakage of steam over the tips of both the rotating and the guide vanes. That this leakage may not be unduly large, the drop in pressure is made small from stage to stage. The leakage at the high-pressure end gradually expanding does some work in the succeeding stages and becomes relatively small at the low-pressure end. Clearances, however, in this type of turbine are all-important and, other things being equal, that reaction turbine showing the best economy will be the one with the smallest radial clearance. To make the leakage in the high-pressure stages as small as possible, the vanes should be made as long as convenient, about five per cent of the diameter of the rotor being considered a minimum.

In turbines of the Rateau type, there is a diaphragm separating the different stages. This extends close to the shaft, permitting leakage only through the small annular space between shaft and bushing, a comparatively unimportant matter, as these clearances may be made very small. Besides this, carbon or labyrinth packing may be provided, which will make the leakage from stage to stage almost negligible. Moreover, there is no tendency to leak over the tips of the running vanes, because the pressure is the same on both sides of them.

In the reaction type of turbine, however, there is no diaphragm from stage to stage and, instead of each rotating wheel being a separate element, it is customary to fasten the vanes in rows to the rotating drum or cylinder. In this style of machine one set of guide vanes and one set of revolving vanes constitute a stage, and it can readily be seen that the opportunity for leakage between the tips of the stationary vanes and the drum of the rotor is very much greater

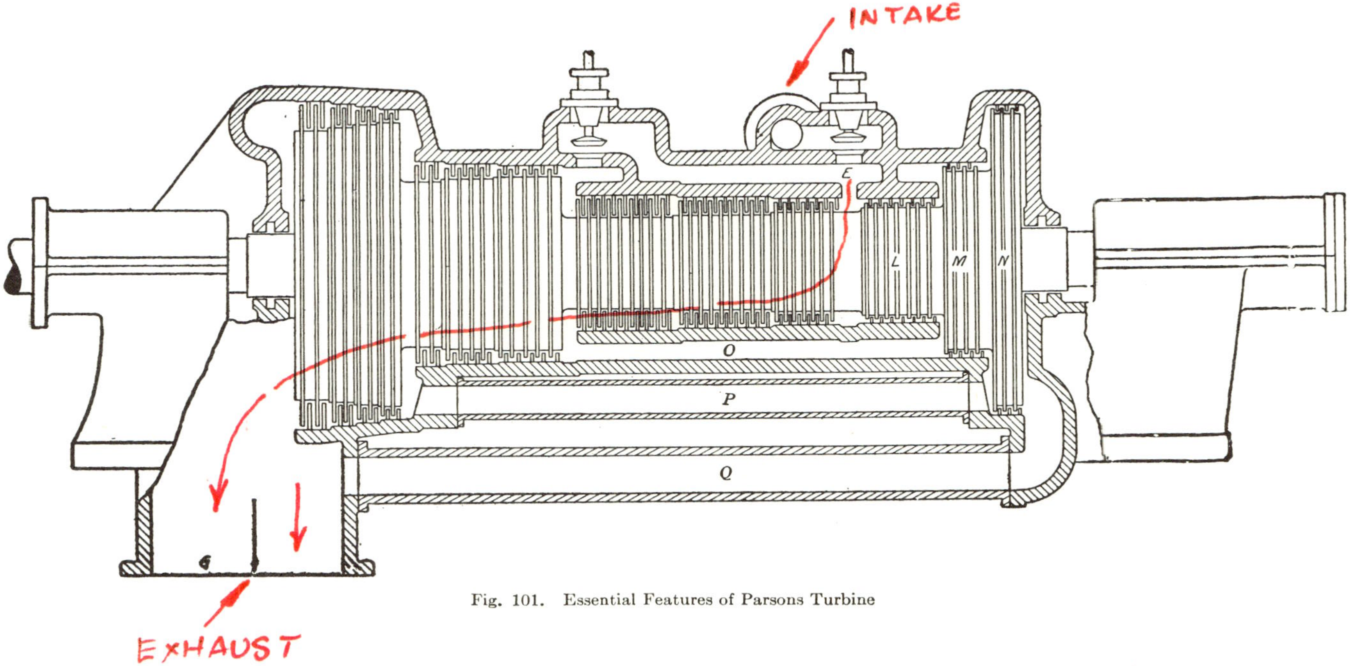

Fig. 101. Essential Features of Parsons Turbine

1. VANE LENGHT REMAINS SAME, HUB DIAMETER INCREASES

than in the Rateau type, the annular space in the two types being proportional to the diameters of shafting and drum. The tendency to leak over the tips of the running vanes is even greater, because of a greater diameter. However, friction losses, which are approximately proportioned to the square of the steam velocity with reference to the vanes, will be very much less in the reaction type of turbine than in the impulse type, because the steam velocities are comparatively low.

Parsons Turbine. The Honorable Charles A. Parsons of England is responsible for the successful development of the reaction type of turbine. His first turbine, made in 1884, was of 10 h.p., 18,000 r.p.m., and when running non-condensing with 92 pounds of steam pressure by the gage, it is claimed that only 25 pounds of steam per brake horsepower per hour were consumed. In 1888, a 50-h.p. turbine at 7000 r.p.m. was constructed, and soon after a 200-h.p at 4000 r.p.m. showed good economy. It must be remembered that the chief problem of the turbine designer has been to reduce rotative speeds without material sacrifice in economy.

Parsons turbines are manufactured in the United States by the Westinghouse Machine Company of Pittsburgh, and by the Allis-Chalmers Company of Milwaukee, Wisconsin, the former acquiring the right to manufacture in 1885, and putting the first turbine on the market three years later. For marine purposes, licenses to build Parsons turbines have been issued to several firms.

The essential features of the Parsons turbine are clearly illustrated by the cross section shown in Fig. 101. Steam enters at E and, in passing through the annular space between the cylinder walls and rotating elements, gradually expands in volume until it exhausts at G. The rotor is usually built in three different diameters to facilitate mechanical construction and to avoid excessively small and excessively large vanes. It is thus possible to use a large number of vanes of the same size. When the length of vane would otherwise become too great, the same passage area may be provided by shortening the vane and increasing the diameter of the drum. While, theoretically, the passage area of the vanes should gradually increase, it is found in practice that without any detrimental effect in economy several rows of vanes may be made of the same height, and thus the areas will increase step by step instead of in a gradual curve.

Since the pressure of steam is greater on the steam side of the vanes than on the exhaust side, there will result an end thrust which must in some way be balanced. This thrust, due to the static pressure of the steam, is augmented by the thrust on the vanes due to the impact and reaction of the steam in passing through them. To balance these thrusts, the balancing pistons shown at *L*, *M*, and *N*, Fig. 101, are provided. These are connected to the steam and exhaust spaces by the passages *O*, *P*, and *Q*, so that the pressure can be readily balanced. It is, of course, necessary to provide some sort of thrust block to meet the requirements of varying conditions, but it need not be large, and it serves in general as an adjustment bearing

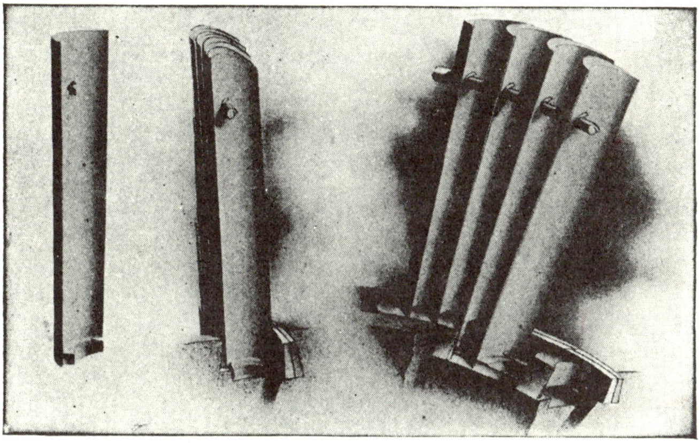

Fig. 102. View of Reaction Blading, Showing Method of Mounting
Courtesy of Westinghouse Machine Company, East Pittsburgh, Pennsylvania

to keep the rotor in correct alignment, and is so arranged that the longitudinal position of the rotor may be slightly changed if desired. A mark on this block shows when the rotor is in its correct position.

Casing. The casing is made with diameters to accommodate the various sizes of drums and blading on the rotor. On the inside of the casing are the rings of fixed vanes or guides which fit between the rings of rotating vanes on the drum. The casing is divided horizontally, so that by lifting the cover all working parts are exposed. In very large turbines the cover slides over four graduated guides, one at each corner, so that in lifting it the engineer can readily see that it is always kept horizontal while being moved, in order to avoid binding or injuring the blading.

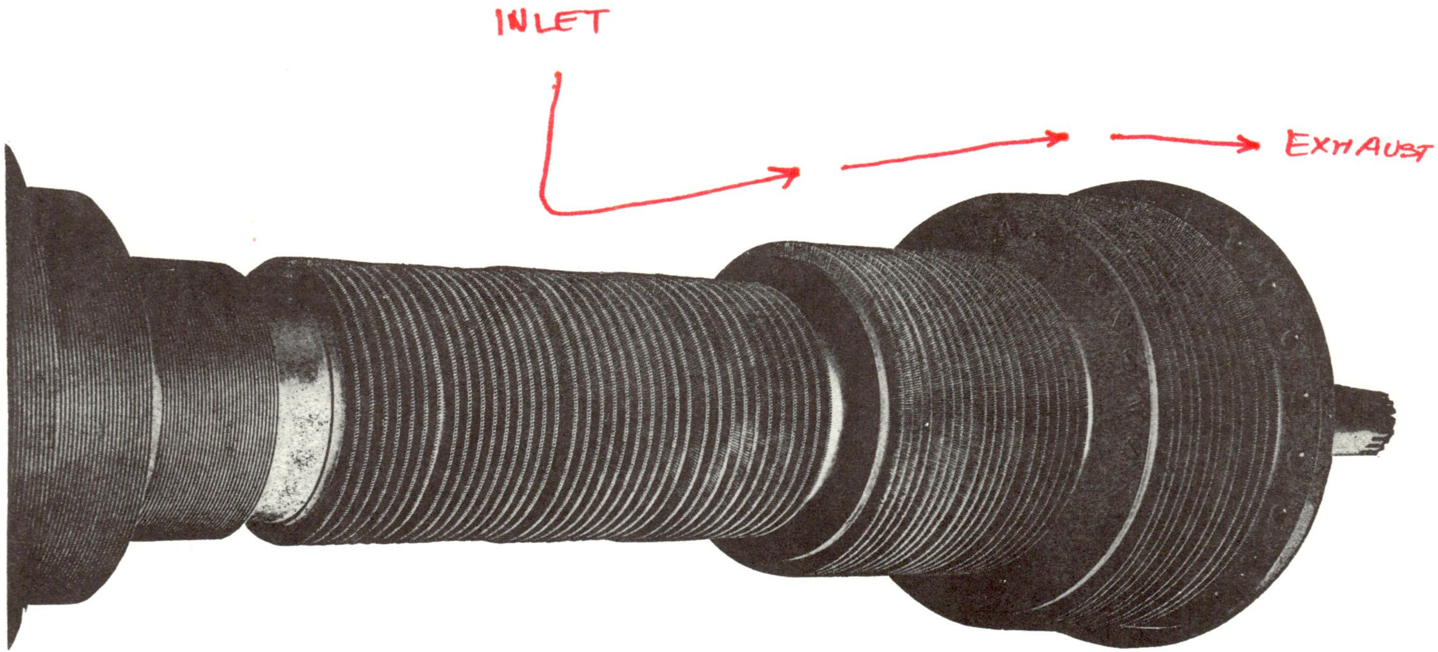

Fig. 103. Rotor of Westinghouse-Parsons Turbine with Balance Pistons

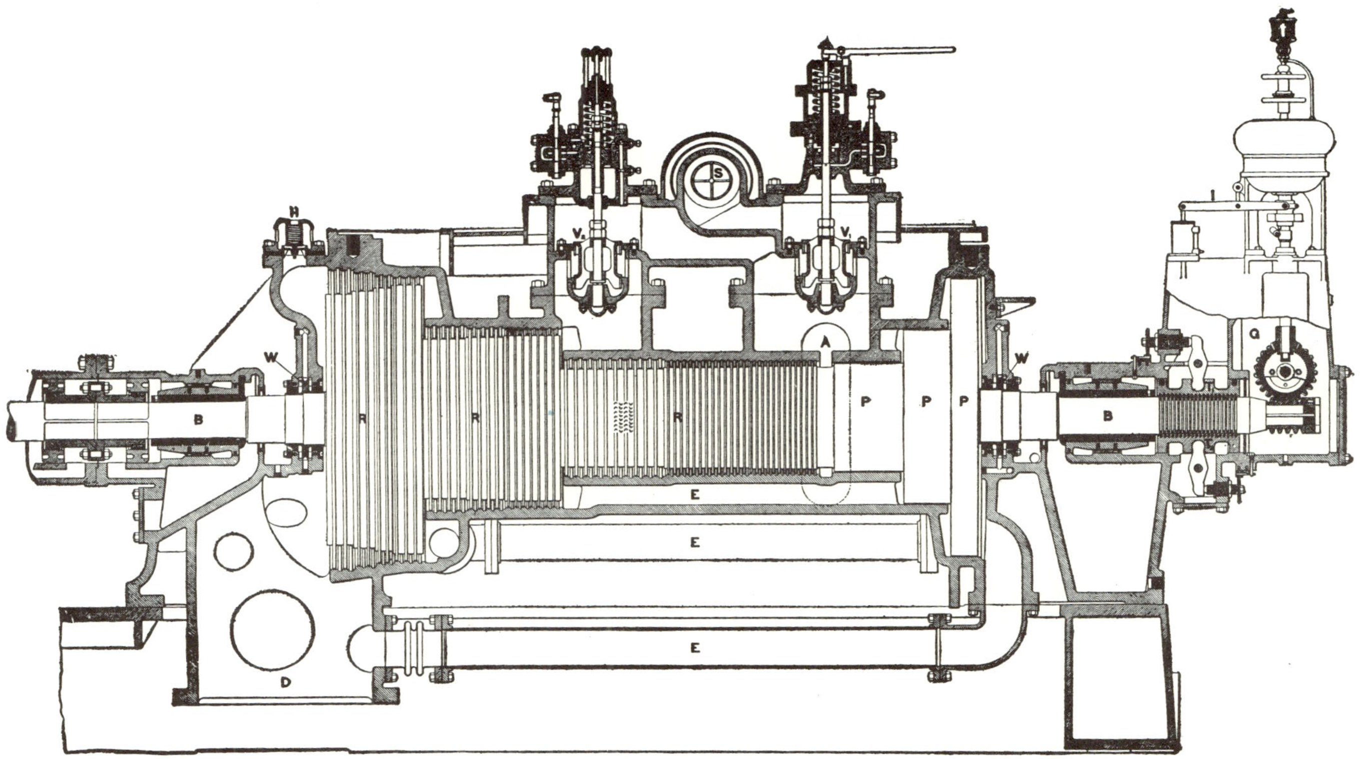

Fig. 104. Section of Westinghouse-Parsons Turbine

Vanes. The vanes are made of suitable material, drawn to the proper section, heat treated, and cut to length. The root ends are slightly upset and a small hook or shoulder formed, as shown in Fig. 102. The grooves in the rotor are of dovetail section and have a small auxiliary groove of rectangular section in the bottom. The steel distance pieces fit snugly in the dovetailed grooves and the shoulders on the lower end of the blades fit down into the rectangular groove and hook under the distance pieces. This is a very strong

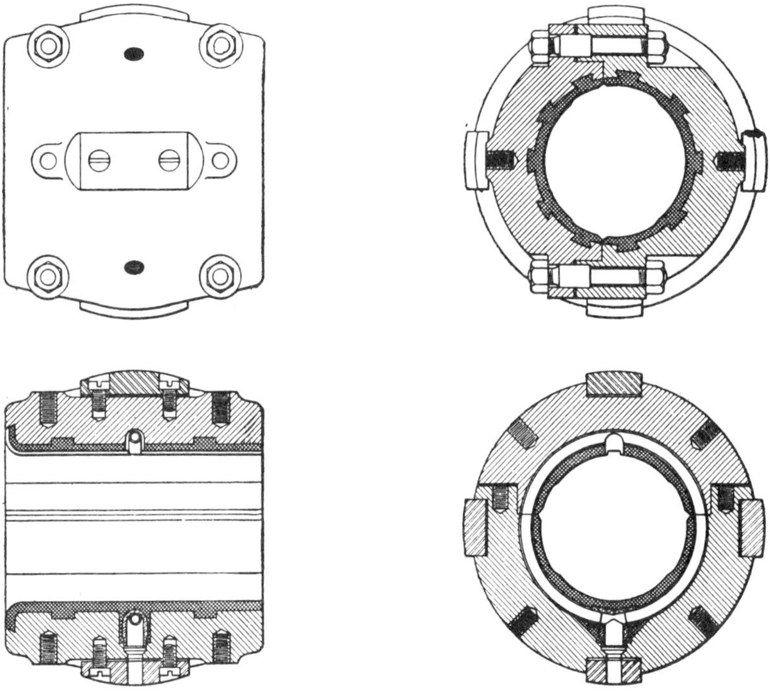

Fig. 105. Bearings of Westinghouse-Parsons Turbine

construction, as the blades could not be pulled out without actually shearing the metal at the shoulder. To stiffen the outer ends of the blades, and to maintain a uniform spacing, a wire lacing is threaded through openings near their outer edge, twisted between the adjacent blades, and soldered into position.

The Westinghouse Machine Company use a cold-drawn wire lacing or lashing of comma-shaped cross section. This is threaded through similarly shaped holes near the tips of the blades, and then the tail of the comma is bent over between the blades, holding them firmly together, as shown in Fig. 102.

Rotor. The rotor, as before noted, consists of a drum of three diameters. This drum is usually built up of a hollow steel casting fixed to some form of spider. The solid rotor would be prohibitively heavy. Fig. 103 shows the Westinghouse-Parsons rotor with balance pistons. Fig. 104 shows this turbine in cross section.

Bearing. The rotor being long and heavy, some sort of flexible or self-aligning bearing is desirable, if not absolutely necessary. Flexibility is provided in the De Laval turbine by means of the long, slender shaft, and in Westinghouse turbines, running at more than 3000 revolutions per minute, by a nest of loosely fitting concentric bronze bushings, with sufficient clearance to permit a continuous oil film between each two bushings. This is intended to form a cushion, permitting a certain amount of vibration in the shaft, yet restraining it within very narrow limits.

It has been found by experience that in the large and slower running turbines this so-called flexible bearing is not necessary. For these, the Westinghouse Machine Company have now adopted the style of bearing shown in Fig. 105. Turbines with short rotating shafts would not require a special form of bearing.

Allis-Chalmers Turbine. The turbine built by the Allis-Chalmers Company of Milwaukee is of the Parsons type with a few special modifications. Fig. 106 is a longitudinal section through a turbine showing all essentials and omitting minor details. Steam enters from pipe *C* after passing through the main regulating valve, which is under the control of the governor. Steam enters the cylinder through passage *E* and, turning to the left, passes through the alternate stationary and moving blades, finally passing to the condenser through connection *G*. The balancing pistons are shown at *L*, *M*, and *N*. In the large sizes the piston *N* is placed at the other end of the spindle; this reduces the size of the piston and makes it much stiffer, as it is backed up by the body of the spindle. The cylinder is divided longitudinally into three sections, the end pieces being cast separately; it is also divided horizontally in its central plane.

In the larger diameters the blade rings are made separate from the body of the spindle, and are provided with a taper fit and pressed on. The balance pistons are constructed in the same way.

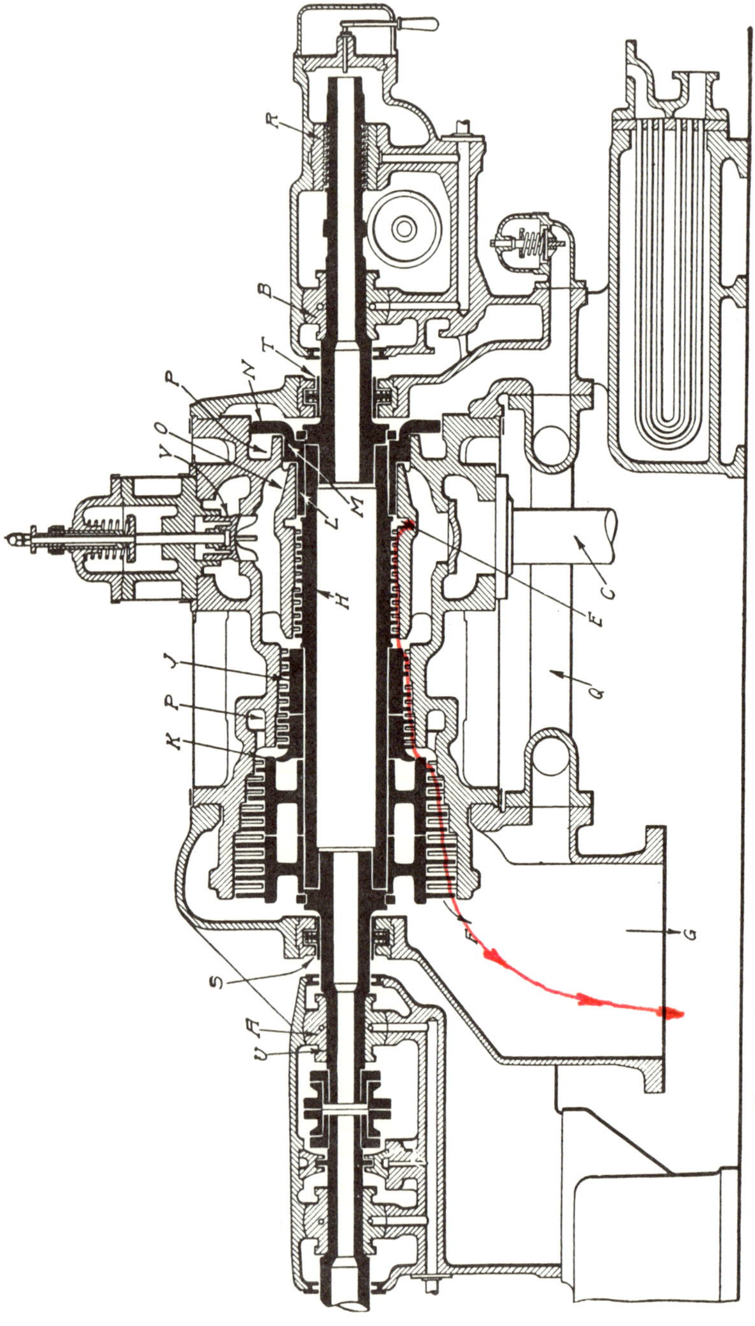

Fig. 106. Longitudinal Section through Allis-Chalmers Steam Turbine—Horizontal Parsons or Reaction Type
Courtesy of Allis-Chalmers Manufacturing Company, Milwaukee, Wisconsin

The bearings are of the ball-and-socket type with spherical seats. The oil is supplied to each of the four bearings by means of pipes located in the lower half of the casing. It passes through the bearing shells and is admitted to the journal at the middle of the top shell. The chief purpose of the collar bearing at the end of the

Fig. 107. Method of Fastening Blades of Allis-Chalmers Turbine

spindle is to give an accurate axial adjustment. It will also take up any small unbalanced thrust which may occur.

In all reaction turbines, expansion of the casing is a troublesome feature. As this is due to the varying temperatures, there is an appreciable difference in the endwise expansion of the spindle and the casing. The high-pressure end of the spindle is held by a collar bearing, and the difference in expansion is taken up at the low-pressure end. The labyrinth packing employed at the high-pressure end has small axial and much radial clearance, while the labyrinth

Fig. 108. Half-Ring Spindle Blading with Blades Laced in the Middle
Courtesy of Allis-Chalmers Manufacturing Company, Milwaukee, Wisconsin

packing of the balance piston at the low-pressure end may have small radial clearance but must have large axial clearance to provide for the difference in expansion. The Allis-Chalmers Company claim that this type of construction permits smaller working clearances in high-pressure and intermediate pistons.

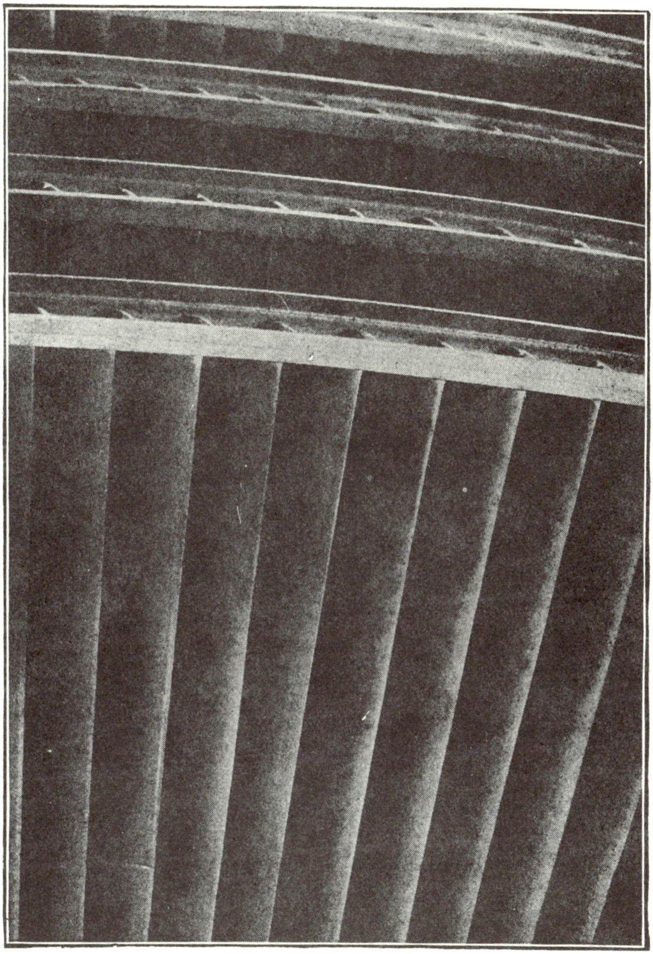

Fig. 109. Sets of Blades Assembled
Courtesy of Allis-Chalmers Company, Milwaukee Wisconsin

Blading. The general shape of vanes of the Allis-Chalmers turbine does not differ from that of other reaction turbines, but the method of securing the blade to the casing and to the rotor is different from that adopted in the ordinary Parsons type. Each blade is so formed that, at its root, it is of an angular, dovetail shape, and has a small projection at its tip. To hold the roots of the blade firmly there is a foundation ring *A*, Fig. 107, which, after being formed to a circle of the proper diameter, has slots cut in it by a special milling machine, these slots being so shaped as to receive the roots of the blade. They are at the same time accurately spaced and so cut as to give the required angle to the blades. To protect the tips of the blades and to bind them together in a substantial manner, a channel-shaped shroud ring *B*, Fig. 107, is fitted. The small projections on the tips of the blade fit through holes cut in this shroud ring and are riveted over. These rings, *A* and *B*, cover half the circumference and the blades are assembled on them, as shown in Fig. 108, before being put onto the rotor or casing. These foundation rings are of dove-tail shape in cross section, and are inserted into corresponding grooves in the turbine casing and spindle, in which they are firmly held by key pieces. These key pieces are driven into place and upset, so as to fill a small undercut, shown at *C* in Fig. 107, thus securely locking them into place. This construction is applied to all blades of whatever length.

The flanges of the channel-shaped shroud rings are made thin, so that, in case of contact with the casing from any accidental cause, no dangerous results are likely to follow, the accidental touch merely causing a slight wearing away of the flanges of the shroud without excessive heating. It is claimed that this shape of shroud ring acts in a measure like a labyrinth packing, retarding appreciably the leakage of steam. Fig. 109 shows a number of sets of blades as assemb'ed.

COMBINED IMPULSE AND REACTION TURBINES

A feature of recent development in turbines of large powers has been the construction of combined impulse and reaction machines. The Westinghouse Machine Company was a pioneer company in the development of this type of turbine. The combination consists of a Curtis wheel at the high-pressure end and Parsons blading at the low-pressure end. The Curtis wheel extracts from 20 to 50 per cent of the total expansion energy and thus replaces a large number of rows of Parsons blading at the point of their lowest efficiency. The effect is to greatly shorten the rotor and avoid some serious mechanical difficulties which would be met in large power high-pressure turbines with pure Parsons blading. Fig. 110 shows in diagrammatic form a section through a single-flow turbine of this class. Steam enters the nozzle chamber and is expanded in the nozzles and discharged against a portion of the periphery of the

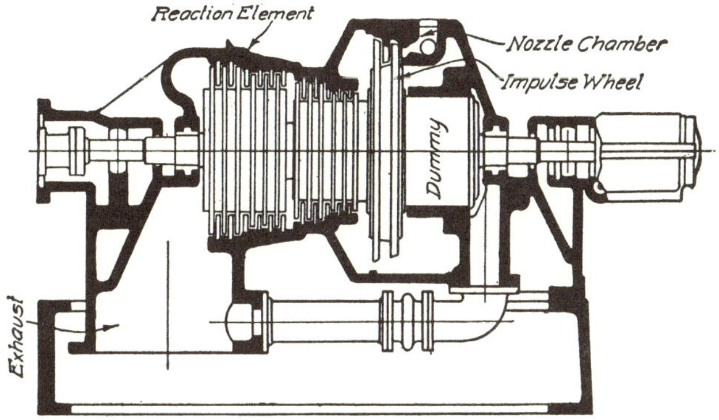

Fig. 110. Section of Combination Impulse and Reaction Single-Flow Turbine
Courtesy of Westinghouse Machine Company, East Pittsburgh, Pennsylvania

impulse wheel. The intermediate- and low-pressure stages are the same as those of a pure Parsons turbine. When the steam enters the cylinder, its pressure and, therefore, its temperature have been much reduced, thus subjecting the cylinder to a small temperature difference only.

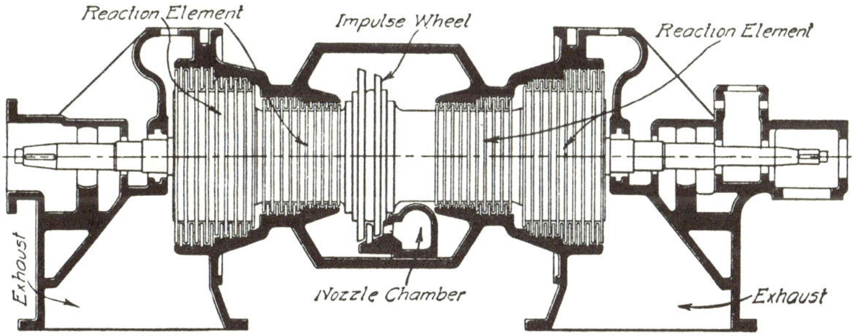

Fig. 111. Section of Double-Flow Steam Turbine
Courtesy of Westinghouse Machine Company, East Pittsburgh, Pennsylvania

Double-Flow Turbine. The capacity of a turbine depends on the weight of steam passed per unit of time; this, in turn, depends on the velocity and the height of the blades. For a given rotative speed the mean diameter of the blade ring is determined by the allowable stress due to centrifugal force, and there is a practical limit to the height of the blades. In the double-flow turbine, Fig. 111, the capacity is doubled by dividing the current of steam after it has passed through the impulse wheel. This type of machine is

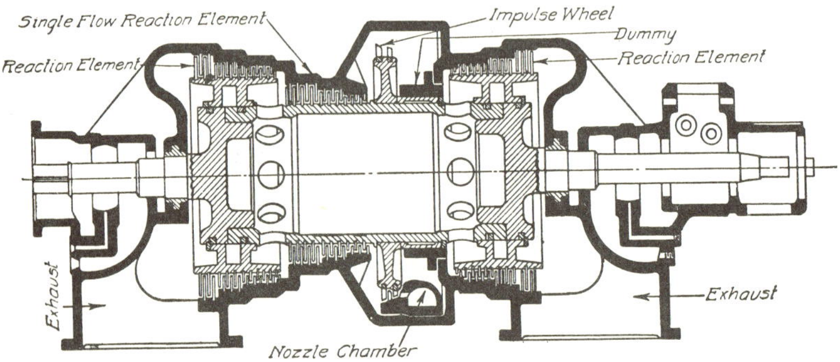

Fig. 112. Section of Semi-Double-Flow Steam Turbine
Courtesy of Westinghouse Machine Company, East Pittsburgh, Pennsylvania

therefore especially adapted for the largest capacities. An incidental advantage of this construction is that the use of dummy or balancing pistons is entirely obviated.

Fig. 113. Rotor of Semi-Double-Flow Steam Turbine
Courtesy of Westinghouse Machine Company, East Pittsburgh, Pennsylvania

In the semi-double-flow turbine, Fig. 112, the steam, after passing through the impulse wheel, passes through a single section of intermediate-stage Parsons blading. It then divides, half of the steam flowing through the drum to half of a double-flow low-pressure reaction blading and then into a double-exhaust opening at either end of the turbine. This construction is now used by the Westinghouse Machine Company for alternating-current turbogenerators in capacities of from 3000 to 10,000 kilowatts, at speeds of 1500 to 1800 revolutions per minute. Fig. 113 shows the construction of the rotor of a semi-double-flow turbine. Each of these constructions has its particular field of usefulness, depending on the operating conditions and the speeds for which the turbines are designed.

GOVERNING

In order to regulate the supply of steam in proportion to the load on the steam turbine, there are several devices employed: (1) The pressure of the steam before admission to the steam chest may be controlled by means of a throttling governor. (2) In the case of impulse turbines with a series of nozzles, each may, if desired, be connected to an independent valve under the control of the governor, so that the number of open nozzles and the extent of the steam belt acting on the turbine may be varied to conform to the load, the steam being admitted to the nozzles at full pressure. (3) Steam may be admitted to the turbine at full pressure at intervals of long or short duration. Any of these devices may be arranged to take care of an ordinary amount of overload. For very large overloads, a by-pass valve is generally provided, which may be controlled either by hand or by the governor, admitting high-pressure steam into one of the lower stages of the turbine.

Of these devices, the second evidently is not applicable to the reaction turbine or to an impulse turbine taking steam around the entire periphery. The other methods of governing might be used for any form of turbine.

Throttling. The simplest governor is undoubtedly the throttle type, but throttling is not an economical way of regulating power; for, by the throttling process, the steam expands somewhat to a lower pressure without doing useful work. The work done goes to superheat the steam and, although no heat units can be destroyed, less can be recovered from a pound of the throttled steam than from a pound at the initial pressure. The smaller the load, and consequently the more the throttling, the greater this loss will be. Any form of turbine, therefore, taking steam about its entire periphery and fitted with a throttling governor, would probably be relatively uneconomical at light loads.

It is not economical to throttle the steam pressure before admission to a set of properly designed nozzles. Nozzles are designed for a definite steam pressure and, if this is varied, the efficiency is bound to

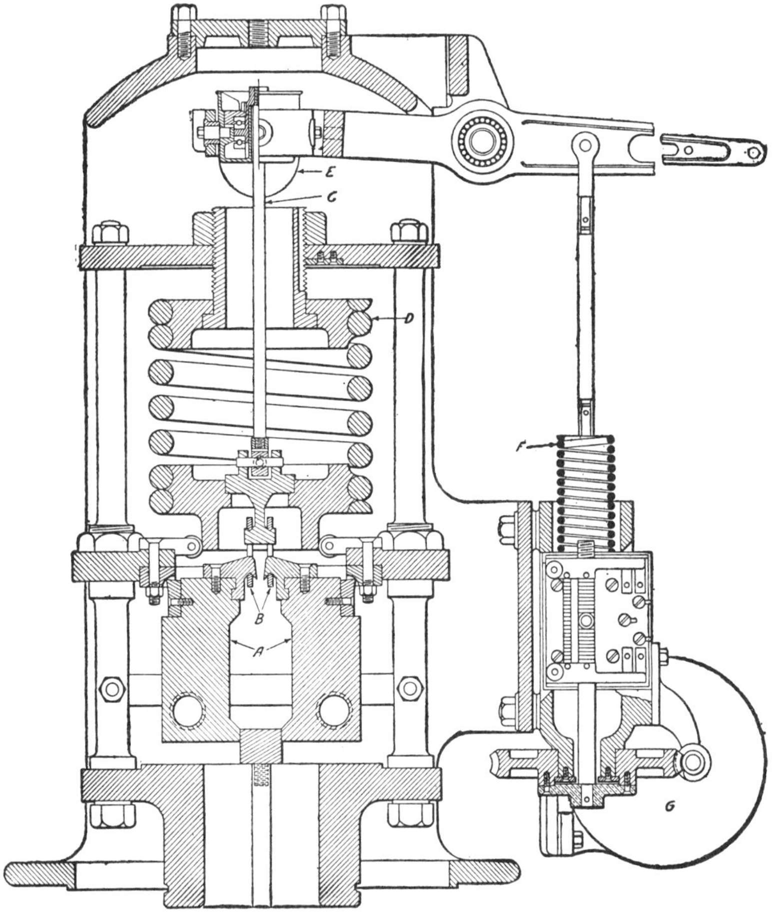

Fig. 114. Sectional View of Governor for Varying Number of Nozzles Opened

fall off. Consequently, there is a twofold loss due to throttling in a turbine of this type. In turbines like the De Laval, supplied with several nozzles, any of which may be opened or closed by hand, it can readily be seen that if there are, for instance, four nozzles, three of which are opened under ordinary loads, then a $33\frac{1}{3}$ per cent or even

$66\frac{2}{3}$ per cent under-load may be taken care of by hand regulation. A throttling governor on such a turbine would necessarily regulate only between these limits and might act only upon one nozzle. On account of its simplicity, the throttling type of governor has been largely used and is practically always used on the smaller turbines, where it has given results that are satisfactory.

Varying Number of Open Nozzles. Turbines employing partial admission through a group of nozzles like those of the Curtis turbine may regulate the steam supply by providing an independent valve for each nozzle. This valve, being under the control of the governor, may be opened or closed as needed. In the Curtis turbine, steam is admitted through a series of valves, the number of which depends upon the capacity of the machine. The valves are arranged to open successively, two-thirds of them being open at full load. The action of the valves is so regulated that they are either fully open or fully closed. Any increasing load is taken care of by the opening of an additional valve, this valve closing when the load falls off. Very wide variations in load can thus be carried with little effect on steam economy. The friction, of course, is a very much larger per cent of small loads, so that the steam consumption at small loads would appear very much larger even though it worked with the same absolute efficiency as at high loads.

These valves are controlled from the governor either by means of some electrical device, by a direct mechanical control, or by hydraulic pressure. The power to operate one of these valves is necessarily so great that it cannot be moved directly by the governor, but the governor can operate a small pilot valve, which in turn may set some mechanism in motion powerful enough to do the work.

The governor itself is of the centrifugal type, its action depending on the balance between the forces exerted by the springs and the centrifugal forces of the revolving weights. Fig. 114 shows a sectional view of this governor. *A* is the revolving weight which, acting through the knife, edges *B*; may move the rod *C* against the action of the spring *D*. At *E* is a ball-bearing gimbal joint which forms a junction between the revolving mass of the governor and the stationary lever of the governing arm. The governor is provided with the auxiliary spring *F*, which may be used to vary the speed of the turbine. *G* is a small motor, which may be operated from the switchboard to regulate the tension of this spring *F*, thus varying the speed of the turbine. The movement of the governor lever is transmitted through a connecting rod and lever to the pilot valve of an hydraulic cylinder. To provide for overloads greater than 50 per cent, an auxiliary set of admission nozzles is provided in the second stage of the turbine which may take full-pressure steam if needed.

Varying Time of Admission. The third method of governing, that by varying the time of admission, is almost invariably used on

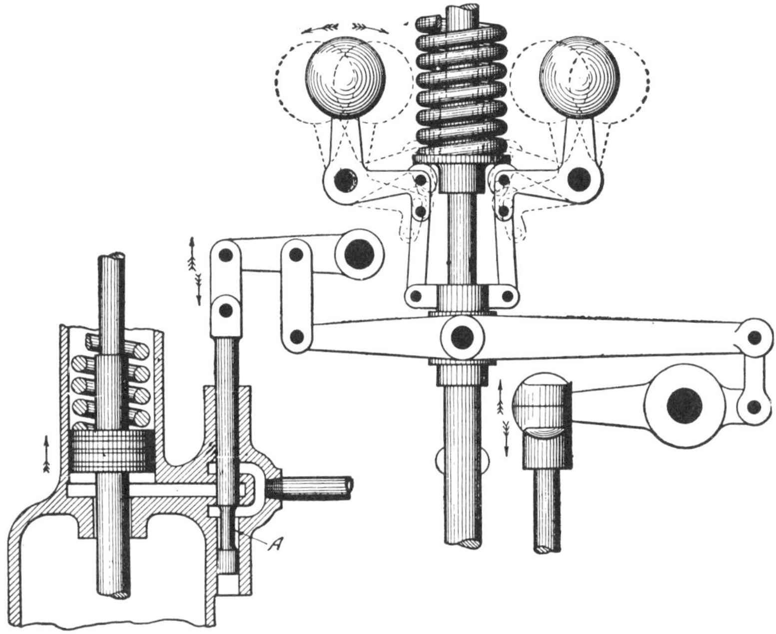

Fig. 115. Turbine Governor for Varying Time of Admission

turbines of the Parsons type. An admission of steam occurs about once in every thirty revolutions at approximately full load. The pilot valve is continually oscillating, thus preventing any liability of sticking, but its period of oscillating is varied directly by the governor. Hence, steam enters the turbine in puffs, the duration of which depends upon the load; at slight overloads the valve will be constantly open. One disadvantage of this method of governing is that there may be an opportunity for the turbine to cool between successive blasts of steam, thereby causing initial condensation at moderate and light loads. A by-pass valve, under the direct control of the governor, is provided on Parsons turbines. For considerable overload this valve will open, letting steam into the intermediate stages.

The details of the mechanism for controlling steam-turbine governors are not essentially different from those used in reciprocating engines, except such as would be required by the greater speed of revolution. In turbines of large size, the valve cannot be operated directly by the governor, but must always be moved by means of pressure admitted through a small pilot valve. Fig. 115 shows the diagrammatic arrangement of the Westinghouse-Parsons governing gear. *A* is the pilot valve under the direct influence of the governor which admits steam to the large piston, shown at the left. This piston has pressure enough upon it to operate the admission valves.

INDEX.

INDEX

W

Z

JET ENGINE

1. INCREASE IN VELOCITY OR INCREASE IN PRESSURE
2. 32° BEND ON BLADES FOR IMPULSE